Geostatistical Analysis of Compositional Data

International Association for Mathematical Geology
STUDIES IN MATHEMATICAL GEOLOGY

1. William B. Size, Editor
   Use and Abuse of Statistical Methods in the Earth Sciences
2. Lawrence J. Drew
   Oil and Gas Forecasting: Reflections of a Petroleum Geologist
3. Ricardo A. Olea, Editor
   Geostatistical Glossary and Multilingual Dictionary
4. Regina L. Hunter and C. John Mann, Editors
   Techniques for Determining Probabilities of Geologic Events and Processes
5. John C. Davis and Ute Christina Herzfeld, Editors
   Computers in Geology—25 Years of Progress
6. George Christakos
   Modern Spatiotemporal Geostatistics
7. Vera Pawlowsky-Glahn and Ricardo Olea
   Geostatistical Analysis of Compositional Data

Geostatistical Analysis of Compositional Data

Vera Pawlowsky-Glahn
Ricardo A. Olea

OXFORD
UNIVERSITY PRESS

2004

OXFORD
UNIVERSITY PRESS

Oxford New York
Auckland Bangkok Buenos Aires Cape Town Chennai
Dar es Salaam Delhi Hong Kong Istanbul Karachi Kolkata
Kuala Lumpur Madrid Melbourne Mexico City Mumbai Nairobi
São Paulo Shanghai Taipei Tokyo Toronto

Published by Oxford University Press, Inc.
198 Madison Avenue, New York, New York 10016

www.oup.com

Library of Congress Cataloging-in-Publication Data
Pawlowsky-Glahn, Vera.
Geostatistical analysis of compositional data / Vera Pawlowsky-Glahn, Ricardo A. Olea.
p. cm. — (Studies in mathematical geology ; 7)
Includes bibliographical references and index.
ISBN 0-19-517166-7
1. Geology—Statistical methods. 2. Multivariate analysis. 3. Kriging.
I. Olea, R. A. (Ricardo A.) II. Title. III. Series.
QE33.2.M3P39 2003
551'.072—dc21 2003058014

9 8 7 6 5 4 3 2 1

Printed in the United States of America
on acid-free paper

Preface

In September 1982 when I decided to enroll at the Free University of Berlin for my graduate studies, the interest expressed by Wolfdietrich Skala in compositional data in a framework relevant to geology led me to devote my dissertation to the study of this topic. I started to go through the literature—mainly publications by Felix Chayes, but also by many other authors; the *Journal of the International Association for Mathematical Geology* became my daily companion.

I grew more and more interested in spatial statistics and began to wonder if there were any problem in regionalized compositional data similar to spurious correlation. I searched for months through papers and books on geostatistics, and finally, not finding an answer to my question, I wrote a paper on *spurious spatial correlation* and presented it at a meeting in Fontainebleau in June 1984. I still remember how nervous I was during my presentation, not only because it was the first time for me to speak in front of an international audience, not only because it was in French, a language I knew only a little bit from school, but mainly because I expected somebody to say: *Haven't you seen that article or that book? There you can find the answer. What you are talking about has been known for a long time!* When I finished speaking, there was just silence. I know now that this is the case in most meetings, but at that moment I felt like falling into a deep hole. Finally Georges Matheron said something like: "There are certainly problems in the straight application of geostatistics to compositional data." That was all I wanted to hear. This was the beginning of my pursuit of a fascinating topic that has dominated my professional interest ever since.

More or less at the same time, as I continued to read *Mathematical Geology*, I came across one of the first papers by John Aitchison that treated compositional data without regard to geographical location. I realized that here was the answer to many of my questions, that integrating Aitchison's approach to Matheron's theory was the way to proceed. I was lucky enough to get a grant from the DAAD (*Deutsche Akademische Austauschdienst*) to visit John Aitchison in Hong Kong in 1985, shortly before he finished his book on the statistical analysis of compositional data. I went there for three months and was able to work with a preprint of the book. The visit was a magnificent learning experience charmingly complemented by my familiarization with a fascinating new culture.

I defended my dissertation successfully in 1986 and shortly thereafter returned to Spain to begin teaching in my native Barcelona. The next transcendental encounter of my professional career and my work in compositional data took place in 1988 when John C. Davis of the Kansas Geological Survey visited Wolfdietrich Skala and Heinz Burger in Berlin. Davis's interest in compositional data in general and my research in particular resulted in the commissioning of his research assistant, Jo Anne DeGraffenreid, to put her German to use by translating my dissertation into English. After corresponding by post, John Davis proposed that I give the translation a final review for publication as a monograph and generously invited me to visit him in Lawrence, Kansas—a visit that did not take place until 1991.

Reviewing Jo Anne's translation was an easy task, but the desire to revise, update, and expand what originally had been my dissertation proved to be a challenge utterly larger than expected, one that took several additional visits to Lawrence and more time than I ever imagined.

So here is the publication of a much delayed work on compositional data as it stood at the conclusion of the second millennium. Although the subject remains open-ended, this seems coincidentally to be an opportune moment to publish the findings before moving on to further developments.

Publication of a book is rarely the exclusive effort of a single individual. As already mentioned, Wolfdietrich Skala, Heinz Burger, and particularly John Aitchison have been extraordinarily supportive of the research summarized here. I am grateful to John Davis and the Kansas Geological Survey for support that was instrumental in advancing my research. Jo Anne DeGraffenreid, after initiating this monograph with her translation, is now serendipitously bringing it to a conclusion after becoming editor of Studies in Mathematical Geology in 1998. Ricardo Olea of the Kansas Geological Survey, since shortly after my first arrival in Kansas, became so supportive of my work that the only way to properly acknowledge his contribution was to make him coauthor of this monograph. Actually, without his constant pushing and particularly his preparation of the case study that comprises the last chapter, there still would be no end in sight. Sometimes I am tempted to add an "s" to Ricardo's surname in order to extend my appreciation to all the Oleas in and from Lawrence who gave me such a comfortable place in their family during my frequent visits.

Completion of this treatise has taken so much time that I am no longer working at the same place where I started my task. Until September 2000, I was a member of the Department of Applied Mathematics III at the Technical

University of Catalonia (UPC). Perhaps UPC's most significant contribution to this monograph, and what has now become my crusade to advance and promote the study of compositional data, was its support during the organization of an IAMG conference in 1997 in which the main topic of discussion was compositional data. I also am indebted to Juan José Egozcue of UPC for ongoing lengthy and productive discussions. At different stages of my career I have also received the important support of Carles Barceló Vidal and Josep Antoni Martín Fernández, both with the Department of Informatics and Applied Mathematics—my new home at the University of Girona.

To my mother Helga and my daughter Tania, and to all the Pawlowsky & Co. clan, thanks for your patience. You know you are the most important people in my life, although sometimes it may not be apparent.

Vera Pawlowsky-Glahn
Girona, December 2000

Contents

List of Figures

List of Tables

Foreword to the Series

The *Studies in Mathematical Geology* series, issued under the auspices of the International Association for Mathematical Geology, was established in 1984 with the idea of providing an outlet for book-length contributions on topics of special interest to the geomathematical community. I share the hope expressed by Dr. Richard B. McCammon, founding editor of the IAMG monographs, that these volumes will provide valuable working and research tools in all facets of geology, and in all the interdisciplinary branches that look to the Association for leadership in the application and use of mathematics in geological research and technology.

Studies in Mathematical Geology No. 7 deals with the analysis and estimation of compositional data in a spatial setting. This particular topic has not been addressed in book form, despite the importance of compositional data in the Earth sciences. Here, the authors extend the statistical concepts of Aitchison and Matheron to develop a model that allows the spatial dependencies of the variables that constitute a regionalized composition to be investigated without the distortions of spurious spatial correlations.

Finally, though perhaps a bit unorthodox for a series foreword, I must take this opportunity to acknowledge the support of my more-than-able associate editors, Drs. Heinz Burger, Freie Universität Berlin, and Thomas A. Jones, ExxonMobil Upstream Research, USA, and thank them for their help. And for their amicable perseverance during trying times, my thanks are due also to the authors, Vera and Ricardo.

Jo Anne DeGraffenreid, Editor
Lawrence, Kansas, USA

Foreword

Independence is a basic concept underlying much of classical statistics. Variables and events are assumed to be independent, permitting the estimation of probabilities from samples and the testing of results using theoretical distributions. Unfortunately, independence is a property of an idealized universe, and in the "real" world the assumption of independence often is unwarranted. This is especially true in the geosciences and other observational disciplines where controlled experimentation (designed to help achieve independence) is not possible. The two most troublesome aspects of non-independence in geological data are spatial dependencies between observations and the imposition of constraints on variables that must sum to a constant. Almost every measurement in the geological sciences is characterized by the physical location of the observations, and the variables often are expressed in terms of percent, ppm, or other fractional measure. The difficulties of analyzing such data using classical statistical tools have led to development, through collaboration between Earth scientists and statisticians, of special concepts and procedures. The first aspect of non-independence is addressed by *geostatistics*, the second by *compositional data analysis*. This volume addresses the area of intersection of these two methodologies and treats the analysis of compositional variables as they vary in space and time. Like earlier developments, it is a collaboration between a mathematician (VP-G) and an Earth scientist (RAO).

As noted by Pawlowsky in the Preface, the Kansas Geological Survey has played a role in nurturing this book and bringing it to final publication—adding it to a long and distinguished list of contributions to mathematical geology fostered by the KGS. However, the Survey's interests in this latest work have not been purely altruistic, because it was recognized that research into the spatial aspects of compositional data provided solutions to many practical local problems in the Earth sciences, including reservoir characterization, aquifer modeling and environmental investigations. I am very pleased to have been able to contribute in a modest way to the publishing of this volume and look forward to the advancements that will be built upon its foundation.

John C. Davis
The University of Kansas

Geostatistical Analysis of Compositional Data

Chapter 1

Introduction

1.1 Statement of the problem

Geological data, notably geochemical data, often take the form of a regionalized composition. The concept of *regionalized composition* combines the concepts of *composition* and *coregionalization*. A composition, also known in the literature as a *closed array* (Chayes 1962), is a random vector whose components add up to a constant. A coregionalization is a set of two or more regionalized variables defined over the same spatial domain, which is modeled as a realization of a vector random function. Here the term *regionalized composition* is used both for the vector random function used to model a composition and for the realization that we can observe.

A regionalized composition can be, for example, a heavy-mineral suite along a river valley. The minerals are quantitatively determined through frequency counts and represented as percent-proportions of the entire heavy-mineral occurrence. Another example is the set of grades in a lead-copper-zinc deposit. In this instance, all components of each specimen are not quantitatively recorded and the grades are also not expressed as proportions of the whole of the measured components: only a small fraction of the composition in ppm is accounted for in each specimen.

The problem with the statistical analysis of compositions has been stated historically in terms of correlations: the covariances are subject to essential nonstochastic controls, i.e., distortions which are due to the constant-sum constraint. These numerically induced covariances and correlations arise also with regionalized compositions and are called *spurious spatial correlations*. They falsify the picture of the spatial covariance structure and can lead to misinterpretations. This problem arises not only when the whole regionalized composition is analyzed, but also when interest lies only in a subvector.

A second problem, singularity of the covariance matrix of a composition, has generally been considered only from a numerical point of view. Singularity is a direct consequence of the constant-sum constraint and, as in other multivariate methods, it rules out the use of estimation techniques such as cokriging of all components. Numerically the problem can be tackled either by taking generalized inverses or, equivalently, leaving one component out to avoid singularity of the matrices of coefficients. Nevertheless, the sum of estimated values differs frequently from the constant corresponding to the whole composition, although the individual values themselves may be reasonable.

The concept of spurious spatial correlation and the singularity of the covariance matrix function are addressed in Chapter 3; a numerical example is given in Chapter 7. Both problems are actually closely related to the basic assumption that the sample space is unconstrained, an assumption implicit in the statistical analysis of coregionalizations, and to the assumption that the distribution of the estimation error at each node of the sampling region is Gaussian.

Although the usual estimation techniques for coregionalizations, like ordinary cokriging, are assumed to be model independent, the absence of constraints on the sample space actually implies that the components of the coregionalization range freely between minus infinity and plus infinity. In other words, they cannot be constrained to assume values on an interval of the real line. As a consequence, using the kriging variance to compute symmetric confidence intervals may lead to intervals for data measured in percentages or ppm that include negative, and thus impossible, values. If a confidence is assigned to the interval—usually 95% for a confidence interval centered at the estimated component plus or minus two times the standard deviation—then a Gaussian model is assumed, thus closing the circle, as a Gaussian model requires an unconstrained sample space. The result could be even worse if the approach described in Chilès and Delfiner (1999, p. 177) were used. In such a case, the three-sigma rule of Pukelsheim (1994) is recommended to obtain a 95% confidence interval under the assumption that the error distribution is continuous and unimodal, which is certainly correct for an unconstrained sample space.

In addition, in a multivariate context, a confidence region can be defined around the estimated vector which will be a multidimensional ellipsoid. For illustrative purposes, consider three components that form a composition. Valid realizations have to satisfy the constant-sum constraint and thus must be included in the corresponding simplex, which in this case is nothing but a ternary diagram. A three-dimensional ellipsoid centered around a predicted value will include possible realizations only in the intersection of the ternary diagram and the ellipsoid, while realizations outside the intersection will not satisfy the constant-sum constraint. These problems call for a methodology that takes into account the interval-restrictions imposed on the sample space of regionalized compositions.

The objective of the work at hand is first to develop a model that allows the spatial dependencies of variables which constitute a regionalized composition to be investigated without the distortion of spurious spatial correlations and then to make joint estimation and thus cokriging of compositional data possible. The methodology followed consists of extending J. Aitchison's theory of statistical analysis of compositions by combining it with G. Matheron's theory of regionalized variables. Therefore, our basic references are Matheron (1965, 1971) and Aitchison (1986).

1.2 Compositions

As early as 1897, Pearson made known the problem of spurious correlations in a nonregionalized composition. A compilation of the most important characteristics is found in the book on ratio correlation by Chayes (1971). This problem is present in all multivariate statistical evaluation methods based on a correlation or covariance analysis of compositions. Major concerns indicated by published articles deal with triangular and Harker diagrams, dendrograms, principal components analysis, multiple correlation analysis, regression analysis, factor analysis, partial correlation analysis, canonical correlation analysis and cluster analysis. Many authors have considered the problem of correlation in connection with compositions. Previously suggested solutions, however, have proved to be inadequate or impractical. A summary review of most of the theoretical and applied works published before 1985 can be found in Aitchison (1986, pp. 61–62). An analysis of articles published after this date would require a whole chapter devoted to them, as an average of 60 articles per year have appeared dealing with compositional data and showing awareness of the problem with spurious correlation. The references can be found in most frequently consulted publication databases. Our purpose here, however, is to analyze regionalized compositions and not compositions in general. Therefore we have limited our study to aspects related to space-dependent data.

In a series of works published in the 1980s, Aitchison developed an alternative for the statistical analysis of compositional data under certain distributional assumptions. See Aitchison and Shen (1980) and Aitchison (1981a, 1981b, 1982, 1983, 1984a, 1984b, 1985). Attention is drawn here particularly to Aitchison's 1986 monograph on the subject, *The Statistical Analysis of Compositional Data.* Further developments appeared in Aitchison (1989, 1990, 1992, 1997), Aitchison and Thomas (1998), and Aitchison et al. (2000).

The idea underlying Aitchison's proposition is especially appealing due to its simplicity. Let $\mathbf{W} = (W_1 \quad W_2 \quad \cdots \quad W_D)'$ be any nonnegative random vector that is not a composition. From $\mathbf{W}$ a composition can always be derived by dividing the individual components by the sum of the components: $Z_i = W_i/(W_1 + W_2 + \cdots + W_D)$. Let $\mathbf{Z} = (Z_1 \quad Z_2 \quad \cdots \quad Z_D)'$ be the resulting composition. Then, the relationship

$$\frac{W_i}{W_j} = \frac{Z_i}{Z_j} \tag{1.1}$$

is always valid for any indices $i, j = 1, 2, \ldots, D$, provided that W_j and Z_j are not zero. This property emphasizes the fact that compositional data only contain information about relative magnitudes, not absolute magnitudes.

This construction shows a way out of the restrictions imposed on the covariances by the constant-sum constraint, as no correlations induced by it arise among the ratios. The ratios, however, are mathematically difficult to manipulate, hence Aitchison's suggestion to evaluate the logarithmic ratios of the components of a composition rather than statistically evaluating the composition itself. Note that using the logarithms of the ratios defines a new sample space for the data to be analyzed: the initial simplex is projected into the whole real space. Thus, in all those cases where the transformed variables are multivariate normally distributed, all multivariate statistical methods for which this distribution is the basis are applicable. Taking into account that many of the usual methods are robust, i.e., they are not sensitive to weak deviations from multivariate normality, many of the usual methods can be applied even if multivariate normality cannot be assumed in a strict sense. In those cases where multivariate normality must be clearly rejected, other transformations can be applied, such as the multivariate Box-Cox transformation (Andrews et al. 1971; Rayens and Srinivasan 1991; Barceló-Vidal 1996). Obviously, these transformations will not always lead to a satisfactory solution. Other alternatives, e.g., the skew-normal defined by Azzalini and Dalla Valle (1996), still have to be analyzed for their applicability. Nevertheless, this approach brings us back to standard problems in multivariate analysis which suffer from the same lack of easy-to-apply methods.

1.3 Coregionalization

The concept of *coregionalization* derives from the theory of regionalized variables as developed by Matheron (1965). This theory forms the foundation for spatial-structure analysis, evaluation and estimation of multivariate, spatially distributed data and can be found in many publications. We would recommend Matheron (1971) for historical reasons; Journel and Huijbregts (1978) to gain a thorough insight into the undersampled case and the linear model of coregionalization; Isaaks and Srivastava (1989), Cressie (1991), Goovaerts (1997), Wackernagel (1998), Olea (1999), and Chilès and Delfiner (1999) to find a more updated approach; and Myers (1982) for the introduction of a suitable matrix formulation of cokriging which sets the basis for estimation of the full compositional vector.

Essential questions relating to multivariate regionalized data arise in

- investigations of the genesis of deposits,
  - when maximum correlations do not occur between variables measured at the same point, but are shifted (delay phenomena),
  - when nonsymmetrical spatial covariances are present;
- cosimulation of several variables, when not only reproduction of the spatial structure of the individual variables but also of the spatial dependency between variables is important; and
- joint estimation of several variables (cokriging),
  - to improve estimations of measured variables with unequal sampling density,
  - to reduce estimation errors through better utilization of available information,
  - to reduce exploration and laboratory expenditures in the case of highly correlated variables of which one can be measured only with costly procedures.

Theoretical approaches concerning regionalized compositions can be found in works by Pawlowsky (1984, 1986, 1989), Pawlowsky and Burger (1992), Olea et al. (1993), Pawlowsky et al. (1995), and Olea and Pawlowsky-Glahn (1999), while the only publications dealing with time series of compositional data—or continuous proportions—known to the authors are by Quintana and West (1988) and Brunsdon and Smith (1998), based on the additive logistic normal distribution, and by Grunwald et al. (1993), based on the Dirichlet distribution.

Publications related to coregionalizations and cokriging from a theoretical point of view are obviously much more frequent. Some have been mentioned already at the beginning of the present section. Other approaches like that of Journel (1998) are not considered here, as they deal with the inference of one single variable, called the *primary variable*, instead of considering a full model of coregionalization, the model usually required in the case of compositional data.

Only a few case studies concerning coregionalization are available. The main reason for this has been the considerable increase of work—compared to separate kriging of single components of a coregionalization—required to adjust models to a multivariate, spatially distributed data set. The inflexibility of available models and the poor increase in any resulting advantages are other reasons. The publication by Yao and Journel (1998) of a method for automatic modeling of cross-covariance tables using Fast Fourier Transforms is surely a major step forward in solving a great part of these problems. However, in our opinion, one of the reasons for the slight improvement in results has been due to the misuse of available methods as applied to regionalized compositions.

Published examples of applications to regionalized compositions pertain mainly to the undersampled case or, equivalently, to the prediction of one set of variables using complementary information contained in another set. Studies of full coregionalizations are less frequent. The following list contains publications on coregionalization known to us, with each group subdivided into cases where noncompositional data, compositional data or a combination of both types are used.

1. Examples using undersampled or complementary variables.
   (a) Noncompositional data:
   Aboufirassi and Mariño (1984), Carr et al. (1985), Solow and Gorelick (1986), Krajewski (1987), Ahmed and de Marsily (1989), Bonano et al. (1989), Creutin et al. (1989), Pereira et al. (1989), Ribeiro and Muge (1989), Stein and Corsten (1991), Atkinson et al. (1992), Pesti et al. (1993), Helterbrand and Cressie (1994), and Abbaspour et al. (1998).
   (b) Compositional data:
   Borgman and Frahme (1976), Journel and Huijbregts (1978), Bryan and Roghani (1982), McBratney and Webster (1983), Carr and Myers (1984), Myers and Carr (1984), Trangmar et al. (1986), Glass et al. (1988), Bordessoule et al. (1989), Daly et al. (1989),

Smyth and Istok (1989), Pan et al. (1993), Asli and Marcotte (1995), and Abbaspour et al. (1998).

(c) Combination of both:
Guarascio (1976), Davis and Greenes (1983), Unal and Haycocks (1986), Yates and Warrick (1987), Doyen (1988), Goulard (1989), Leenaers et al. (1989), Muge and Cabeçadas (1989), and Goovaerts and Webster (1994).

2. Examples where a full vector (not necessarily constant sum) is analyzed.

   (a) Noncompositional data:
   Carr and McCallister (1985), Clark et al. (1989), Creutin et al. (1989), Stein et al. (1991), ver Hoef and Cressie (1993), and Morisette (1997).

   (b) Compositional data:
   Stattegger (1983), Grunsky and Agterberg (1988), Pawlowsky and Stattegger (1988), Daly et al. (1989), Grunsky (1989), Wackernagel and Butenuth (1989), Royer (1989), Sousa (1989), Bourgault and Marcotte (1991), Grunsky and Agterberg (1992), Pawlowsky and Burger (1992), Olea et al. (1993), Goovaerts and Webster (1994), Pawlowsky et al. (1995), and Yao (1998).

   (c) Combination of both:
   Doyen (1988), Pan (1993), Goovaerts (1994b), and Hendriks et al. (1998).

It has to be pointed out that only Pawlowsky and Stattegger (1988), Royer (1989), Pawlowsky and Burger (1992), and Olea et al. (1993) make explicit reference to the compositional character of the sample. McBratney and Webster (1983) use the logit transformation; Smyth and Istok (1989) and Goovaerts and Webster (1994) use a lognormal approach, which can be justified in these cases because only trace elements are under study (see Section 5.4.2 for further details). An interesting univariate example using model-based geostatistics combined with the lognormal distribution has been published by Journel (1980). Diggle et al. (1998) combine model-based geostatistics with the logit transformation.

1.4 Organization of the book

This monograph is intended as a "state-of-the-art" book rather than as a textbook. It gives the general framework for regionalized compositions as well as for cokriging of a whole vector. The case of cokriging of a single variable as a function of other variables (useful in the undersampled case) is not included.

Chapter 2 introduces basic concepts and definitions for regionalized compositions (abbreviated *r-compositions*) which are mainly direct extensions to the regionalized case of concepts involved in the statistical analysis of compositional data: composition, basis of a composition, subcomposition, amalgamation and partition. Also included are necessary adjustments of the basic hypotheses of the theory of regionalized variables, from stationarity and intrinsic hypothesis to compositions. In addition, the additive logratio (alr) and centered logratio (clr) transformations are referenced. Additive logratio transformations lead to the additive logistic normal distribution [see Aitchison (1986, pp. 113–114) and Aitchison and Shen (1980) for further details].

Chapter 3 presents the definition of the spatial covariance structure for r-compositions. The specifications of the covariance structure for nonregionalized compositions are extended to include the basic tools of spatial-structure analysis: cross-covariances and cross-semivariograms. The goal is to establish the foundation of the analysis of spatial correlation and the independence of the r-compositions. We also intend to determine the most suitable specification, i.e., the most suitable way to express the spatial covariance structure of an r-composition in terms of covariances and cross-covariances. Alternatively in this particular case, our goal is to express spatial covariance structure most suitably in terms of semivariograms and cross-semivariograms for cokriging.

From the specifications of the spatial covariance structure in Chapter 3, expressions of null cross-correlation for r-compositions emerge. These are discussed in Chapter 4. The connections between possible expressions of null cross-correlation are also examined in order to determine the requirements for the systematic analysis of the structure of the spatial dependence between components of an r-composition.

Chapter 5 describes the estimation technique known as cokriging when the r-composition follows a multivariate normal, a lognormal, or an additive logistic normal distribution. The latter two cases are particularly interesting for r-compositions. However, because they build upon the former, the case of the normal distribution is presented first. This chapter also contains a proof for the invariance of the alr cokriging estimator under permutation of the r-composition.

Chapter 6 discusses several aspects that are essential for the practical application and interpretation of the proposed methodology, but that are not essential for the previous theoretical developments. Aspects such as how to deal with zeros in compositional data sets, how to model required cross-covariances, and how to build confidence intervals and confidence regions for estimated values in the original sample space, such as in the simplex with D components (or the D-simplex for short) are considered.

With the aid of a geological example, Chapter 7 illustrates the concepts and procedures that have been introduced. The example draws on data from the Lyons West oil field located in Kansas (USA). On the basis of these data, different methods of estimation are discussed, among them the direct approach (which does not take into account the spatial dependence among considered variables), the basis method (which is only feasible when an external variable is available which can be considered to be the *size* of the composition), and finally the alr approach. These different methods are compared and discussed based on results obtained by analysis of the Lyons West data. In summary, this book discusses:

- A method for spatial correlation analysis of regionalized compositions.
- The possibility of using geostatistics with compositional data. Combined with the agl transformation (the inverse of the alr transformation), alr cokriging is an exact interpolation procedure that also yields a regionalized composition, that is, the interpolated values add to the same constant as do the original values.

Chapter 2

Regionalized compositions

In this chapter we set the rationale for the analysis of regionalized compositions. Required definitions for nonregionalized compositions are extended to vector random functions and necessary concepts from the theory of regionalized variables are related to vector random functions that form a composition. In order to avoid continually repeating references to literature, the reader is referred especially to the works of Matheron (1971) and Aitchison (1986), on which the following developments are based. Here the exposition is very concise; its purpose is basically to introduce terminology and notation. Proofs analogous to those of the nonregionalized case are omitted, for the most part. In general they can be derived directly from the corresponding definitions.

For concepts of matrix algebra required by this work, refer to Kemény (1984) and Golub and Van Loan (1989). There are many excellent textbooks that treat concepts of probability theory and multivariate statistics. We have used mainly the books by Fahrmeir and Hamerle (1984) and Krzanowski (1988), and others have served as complementary bibliography, e.g., Feller (1968), Kendall and Stuart (1979), Kendall et al. (1983), Kres (1983), Stuart and Ord (1987), Johnson et al. (1994), and Kotz et al. (2000). A similar situation holds for the foundations of univariate geostatistics; refer to David (1977), Journel and Huijbregts (1978), Rendu (1978), Clark (1979), Isaaks and Srivastava (1989), Samper-Calvete and Carrera-Ramírez (1990), Cressie (1991), Goovaerts (1997), Chilès and Delfiner (1999), and Olea (1999). Treatments of multivariate geostatistics are found in Matheron (1979), François-Bongarçon (1981), Carr et al. (1985), and Wackernagel (1998). We base our presentation mainly on Journel and Huijbregts (1978) and Deutsch and Journel (1998), but also on Myers (1982), in which the matrix formulation of cokriging is given. Geostatistical terminology conforms, as far as possible, to that found in the *Geostatistical*

Glossary and Multilingual Dictionary, compiled by members of the 1984–1989 IAMG Committee on Geostatistics and edited by R. A. Olea (1991).

Throughout the work, the basic concept to model reality is a vector random function with D components, or D-*part vector random function*

$$\mathbf{Z}(\mathbf{u}) = (\, Z_1(\mathbf{u}) \quad Z_2(\mathbf{u}) \quad \cdots \quad Z_D(\mathbf{u})\,)', \quad \mathbf{u} \in \Omega,$$

where $\Omega \subset \mathbb{R}^\nu$ is a spatial domain and $\mathbb{R}^\nu$ is, in general, either one-, two- or three-dimensional real space. The existence of the appropriate probability space is implicitly assumed. The components of the vector random function are assumed to be real random functions. Recall that in the framework of geostatistics, for a fixed point $\mathbf{u}_n$ of Ω, $\mathbf{Z}(\mathbf{u}_n)$ is a random vector; a set of observations of $\mathbf{Z}(\mathbf{u}_n)$, $\mathbf{u}_n \in \Omega$, $n = 1, 2, \ldots, N$, constitutes a partial realization of the vector random function, and the observation actually made at a fixed point $\mathbf{u}_n$ is a vector of real numbers. In practice only a partial realization of the vector random function is attainable. There is no notational distinction in the following between a vector random function $\mathbf{Z}(\mathbf{u}) = (\, Z_1(\mathbf{u}) \quad Z_2(\mathbf{u}) \quad \cdots \quad Z_D(\mathbf{u})\,)'$, the random vector $\mathbf{Z}(\mathbf{u}_n) = (\, Z_1(\mathbf{u}_n) \quad Z_2(\mathbf{u}_n) \quad \cdots \quad Z_D(\mathbf{u}_n)\,)'$ at a fixed point $\mathbf{u}_n$, and a realization at that fixed point $\mathbf{u}_n$, unless it is not clear from the context. Also, the explicit reference to the size D of the vector random function is omitted, unless necessary. Finally, transformations of vector random functions are required. They are defined, as usual, over the set of possible realizations of the vector random function, with the implicit understanding that a new vector random function is thus defined [for a more detailed discussion see Papoulis (1965)]. The transformations that are used satisfy the general conditions necessary for this to be so.

2.1 First concepts of regionalized compositions

Definition 2.1 *A* regionalized composition *(r-composition for short) is a vector random function* $\mathbf{Z}(\mathbf{u})$ *which at each point* $\mathbf{u}, \mathbf{u} \in \Omega$, *satisfies the following conditions:*

1. *all its components are positive:* $Z_i(\mathbf{u}) > 0$, $i = 1, 2, \ldots, D$;
2. *the sum of all its components is constant and equal to one:*

$$\sum_{i=1}^{D} Z_i(\mathbf{u}) = \mathbf{j}'\mathbf{Z}(\mathbf{u}) = 1, \quad \mathbf{u} \in \Omega,$$

where $\mathbf{j} = (1, 1, \ldots, 1)'$ *is a vector of ones, and the prime stands for transpose.*

At each point $\mathbf{u}$, $\mathbf{u} \in \Omega$, $\mathbf{Z}(\mathbf{u})$ is a random vector which satisfies the constant-sum constraint; as a result, for any fixed $\mathbf{u}$, $\mathbf{u} \in \Omega$, the natural sample space for $\mathbf{Z}(\mathbf{u})$ is the D-simplex embedded in D-dimensional real space $\mathbb{R}^D$, that is, its realizations belong necessarily to the set

$$\mathcal{S}^D = \{\mathbf{Z}(\mathbf{u});\ Z_i(\mathbf{u}) > 0,\ i = 1, 2, \ldots, D;\ \mathbf{j}'\mathbf{Z}(\mathbf{u}) = 1\}. \tag{2.1}$$

Remarks.

1. The case $Z_i(\mathbf{u}) = 0$, for any $i = 1, 2, \ldots, D$, is excluded in order to avoid unnecessarily complicating the presentation. Data sets with many zeros are as cumbersome with compositional data as they are in standard multivariate statistics. Nevertheless, as the problem with zeros appears frequently in compositional data sets and the question concerning the way to approach it is a matter of discussion, we address it more extensively in Section 6.1.

2. The condition $\mathbf{j}'\mathbf{Z}(\mathbf{u}) = 1$ is used, although in practice the sum 100% or some other constant is common. This is done for reasons of clarity because the use of logratios makes a conversion unnecessary inasmuch as the constant can be simplified (see definition 2.10).

3. According to equation (2.1) we have the following subset relationships: $\mathcal{S}^D \subset \mathbb{R}^D_+ \subset \mathbb{R}^D$, in which $\mathbb{R}^D_+$ is that part of D-dimensional real space $\mathbb{R}^D$ which extends along the positive semiaxes, also known as the *positive orthant* of $\mathbb{R}^D$.

4. Although $\mathcal{S}^D$ is the natural sample space of a random composition, and therefore the dependence on the point $\mathbf{u}$ is kept in the notation, note that this implies that all the observations at points $\mathbf{u}_n$, $n = 1, 2, \ldots, N$ will be elements of $\mathcal{S}^D$ as a subset of $\mathbb{R}^D_+$.

2.2 Basis of a regionalized composition

Definition 2.2 *A* regionalized basis *(r-basis for short)* $\mathbf{W}(\mathbf{u})$, $\mathbf{u} \in \Omega \subset \mathbb{R}^n$, *is a vector random function whose components are all measured on the same scale and are positive.*

An r-basis is nothing else but a vector of positive components, the closure of which leads to an r-composition. It has been used implicitly to illustrate the fact that r-compositions only contain information about relative magnitudes in equation (1.1). It is equivalent to the concept of accumulation in mining geostatistics (Journel and Huijbregts 1978, p. 199).

At each point $\mathbf{u}$, $\mathbf{u} \in \Omega$, the natural sample space for a D-part r-basis $\mathbf{W}(\mathbf{u})$ is that part of D-dimensional real space $\mathbb{R}^D$ which extends along the positive semiaxes, in other words, its realizations belong to the positive orthant of D-dimensional real space,

$$\mathbb{R}_+^D = \{\mathbf{W}(\mathbf{u}); W_i(\mathbf{u}) > 0, i = 1, 2, \ldots, D\}.$$

Definition 2.3 *The* closure operator

$$\begin{aligned} \check{C} : \mathbb{R}_+^D &\longrightarrow \mathcal{S}^D \\ \mathbf{W}(\mathbf{u}) &\longmapsto \check{C}(\mathbf{W}(\mathbf{u})) = \frac{\mathbf{W}(\mathbf{u})}{\mathbf{j}'\mathbf{W}(\mathbf{u})} \end{aligned}$$

transforms each r-basis $\mathbf{W}(\mathbf{u})$ *into an r-composition* $\check{C}(\mathbf{W}(\mathbf{u}))$.

Definition 2.4 *Every r-basis* $\mathbf{W}(\mathbf{u})$ *has associated a unique* size

$$t(\mathbf{u}) = \mathbf{j}'\mathbf{W}(\mathbf{u}),$$

that is itself a random function, as well as a unique r-composition $\mathbf{Z}(\mathbf{u}) = \check{C}(\mathbf{W}(\mathbf{u}))$.

Property 2.1 *A one-to-one correspondence between* $\mathbf{W}(\mathbf{u})$ *and* $(t(\mathbf{u}), \mathbf{Z}(\mathbf{u}))$ *is given by:*

$$\begin{aligned} \mathbb{R}_+^D &\longrightarrow \mathbb{R}_+^1 \times \mathcal{S}^D \\ \mathbf{W}(\mathbf{u}) &\longmapsto (t(\mathbf{u}), \mathbf{Z}(\mathbf{u})) = \left(\mathbf{j}'\mathbf{W}(\mathbf{u}), \check{C}(\mathbf{W}(\mathbf{u}))\right); \\ \mathbf{W}(\mathbf{u}) = t(\mathbf{u})\mathbf{Z}(\mathbf{u}) &\longleftarrow (t(\mathbf{u}), \mathbf{Z}(\mathbf{u})). \end{aligned}$$

Property 2.2 *The ratio of two arbitrary components of an r-basis is identical to the ratio of the corresponding components of its associated r-composition; that is, if* $\mathbf{Z}(\mathbf{u}) = \check{C}(\mathbf{W}(\mathbf{u}))$, *then*

$$\frac{W_i(\mathbf{u})}{W_j(\mathbf{u})} = \frac{Z_i(\mathbf{u})}{Z_j(\mathbf{u})}, \quad i, j = 1, 2, \ldots, D.$$

Many examples of r-compositions may be found in the Earth sciences, but they occur most frequently in geochemistry. Thus every reduction of a spatially distributed phenomenon into its components (major elements or trace elements measured in the same units, whether % or ppm) results in an r-composition. On the other hand, it is not easy to find examples of r-bases. At least they are not so "naturally" determined. However, if all components in a rock specimen are measured in percent, for instance, and these measurements are related to the thickness of the rock unit from which the specimen was taken, a new measurement vector—the equivalent thickness of the different components—is given by multiplication of the individual variables by the unit's thickness. This vector forms an r-basis $\mathbf{W}(\mathbf{u})$. The r-composition $\check{C}(\mathbf{W}(\mathbf{u}))$ corresponds, in accordance with the definitions, to the original measurement vector. The random function $t(\mathbf{u})$ which results from the r-basis would in this case be the thickness of the unit. It is important to note that $t(\mathbf{u})$ is assumed to be a random function and not a constant, as otherwise $\mathbf{W}(\mathbf{u})$ would be a rescaled r-composition with all its inherent problems.

It is obvious that an individual r-composition corresponds to each r-basis. The reverse is not true, because for every conceivable "external" variable, as for example the already-mentioned thickness, it is possible to determine an r-basis from an r-composition.

2.3 Regionalized subcompositions

Definition 2.5 *If* $s \subset \{1, 2, \ldots, D\}$, $\mathbf{Z}(\mathbf{u})$ *is an r-composition and* $Z_s(\mathbf{u})$ *is the subvector formed from the subset of components of* $\mathbf{Z}(\mathbf{u})$ *with indices in* s*, then* $\check{C}(Z_s(\mathbf{u}))$ *is a* regionalized subcomposition *or, abbreviated, r-subcomposition.*

Definition 2.6 *A* selecting matrix $\mathbf{S}$ *for* $\mathbf{Z}(\mathbf{u})$ *is a* $(C \times D)$ *matrix* $(1 \leq C \leq D)$ *with* C *elements equal to one, at least one of them in each row, and not more than one in each column. The remaining* $C(D-1)$ *elements are zero.*

Property 2.3 *If* $\mathbf{S}$ *is a selecting matrix, then* $\check{C}(\mathbf{SZ}(\mathbf{u}))$ *is an r-composition with* C *components.*

Consequently, every r-subcomposition can be understood to result from a transformation

$$\begin{aligned} \check{C}\check{S} : \mathcal{S}^D &\longrightarrow \mathcal{S}^C \\ \mathbf{Z}(\mathbf{u}) &\longmapsto \check{C}\check{S}(\mathbf{Z}(\mathbf{u})) = \check{C}(\mathbf{SZ}(\mathbf{u})) = \check{C}(Z_s(\mathbf{u})). \end{aligned}$$

Property 2.4 *The ratio between any two components of an r-subcomposition is identical to the ratio between the corresponding components of the r–composition;* i.e., *if* $\mathbf{s}(\mathbf{u}) = \check{C}(\mathbf{SZ}(\mathbf{u}))$*, then*

$$\frac{s_i(\mathbf{u})}{s_j(\mathbf{u})} = \frac{Z_i(\mathbf{u})}{Z_j(\mathbf{u})}, \quad i, j \in s.$$

Now, for $1 \leq C \leq D$, let

$$\begin{aligned}\mathbf{Z}_{(C)}(\mathbf{u}) &= (\, Z_1(\mathbf{u}) \quad Z_2(\mathbf{u}) \quad \cdots \quad Z_C(\mathbf{u})\,)', \\ \mathbf{Z}^{(C)}(\mathbf{u}) &= (\, Z_{C+1}(\mathbf{u}) \quad Z_{C+2}(\mathbf{u}) \quad \cdots \quad Z_D(\mathbf{u})\,)'.\end{aligned}$$

Property 2.5 *Given an index $C \in \{1, 2, \ldots, D\}$ and an r-composition $\mathbf{Z}(\mathbf{u}) = (Z_{(C)}(\mathbf{u}), \mathbf{Z}^{(C)}(\mathbf{u}))'$, if $\mathbf{s}_1(\mathbf{u}) = \check{C}(Z_{(C)}(\mathbf{u}))$ and $\mathbf{s}_2(\mathbf{u}) = \check{C}(\mathbf{Z}^{(C)}(\mathbf{u}))$ denote the corresponding r-subcompositions, then $\mathbf{Z}(\mathbf{u})$, together with C, explicitly determines $\mathbf{s}_1(\mathbf{u})$ and $\mathbf{s}_2(\mathbf{u})$.*

The reverse is not the case because for each random function $t(\mathbf{u})$, $0 < t(\mathbf{u}) < 1$, $\mathbf{u} \in \Omega$, the r-composition of the form

$$t(\mathbf{u}) \begin{pmatrix} \mathbf{s}_1(\mathbf{u}) \\ 0 \end{pmatrix} + (1 - t(\mathbf{u})) \begin{pmatrix} 0 \\ \mathbf{s}_2(\mathbf{u}) \end{pmatrix} = \begin{pmatrix} t(\mathbf{u}) s_{11}(\mathbf{u}) \\ \vdots \\ t(\mathbf{u}) s_{1C}(\mathbf{u}) \\ (1 - t(\mathbf{u})) s_{21}(\mathbf{u}) \\ \vdots \\ (1 - t(\mathbf{u})) s_{2,D-C}(\mathbf{u}) \end{pmatrix}$$

has $\mathbf{s}_1(\mathbf{u})$ and $\mathbf{s}_2(\mathbf{u})$ as r-subcompositions.

Examples of r-subcompositions occur more frequently in geology than examples of r-compositions. For instance, usually not all components of a rock specimen are measured, rather only a fraction of them. These are then often standardized as percents of the quantitatively recorded portion—for example, as oxide percentages. The partitioning of a spatially distributed measurement vector into major elements and trace elements by geochemical analysis is an example of the breaking down of an r-composition into two r-subcompositions.

2.4 Regionalized amalgamations and partitions

Definition 2.7 *If the D components of an r-composition are separated into C ($C \leq D$) disjoint and exhaustive subsets, and if the components of each of the individual subsets are added, then the resulting r-composition with C components is a* regionalized amalgamation *or, abbreviated, r-amalgamation.*

Definition 2.8 *An* amalgamation matrix $\mathbf{A}$ *is a $C \times D$ ($C \leq D$) matrix with D elements equal to one. Every column contains only one of them, and there is at least one in each row. The remaining $(C-1)D$ elements are equal to zero.*

The matrix

$$\begin{pmatrix} 1 & 0 & 0 & 0 & 0 \\ 0 & 1 & 1 & 0 & 0 \\ 0 & 0 & 0 & 1 & 1 \end{pmatrix}$$

is an example of an amalgamation matrix that transforms an r-composition with five components into an r-composition with three components.

Property 2.6 *If an r-composition* $\mathbf{Z}(\mathbf{u})$, $\mathbf{u} \in \Omega$, *is premultiplied by a* $C \times D$ *amalgamation matrix* $\mathbf{A}$, *then the resulting r-amalgamation*

$$\mathbf{V}(\mathbf{u}) = \mathbf{A}\mathbf{Z}(\mathbf{u})$$

is an r-composition with C *components.*

It follows that every amalgamation can be understood as a transformation:

$$\begin{aligned} \breve{A} : \mathcal{S}^D &\longrightarrow \mathcal{S}^C \\ \mathbf{Z}(\mathbf{u}) &\longmapsto \breve{A}(\mathbf{Z}(\mathbf{u})) = \mathbf{A}\mathbf{Z}(\mathbf{u}) = \mathbf{V}(\mathbf{u}). \end{aligned}$$

Property 2.7 *Every* $D \times D$ *permutation matrix, including the identity matrix, is an amalgamation matrix.*

Definition 2.9 *If the* D *components of an r-composition are divided into* C $(C \leq D)$ *disjoint and exhaustive subsets, then the resulting r-amalgamation, together with the r-subcompositions that emerge from the subsets, form a* regionalized partition of order C.

Let $Z_p(\mathbf{u})$ denote the vector whose components are the elements of the p-th subset, $t_p(\mathbf{u}) = \mathbf{j}' Z_p(\mathbf{u})$ the sum of its components,

$$\mathbf{V}(\mathbf{u}) = (\, t_1(\mathbf{u}) \quad t_2(\mathbf{u}) \quad \cdots \quad t_C(\mathbf{u}) \,)'$$

the associated r-amalgamation, and $\mathbf{s}_i(\mathbf{u}) = \breve{C}(Z_i(\mathbf{u}))$ the corresponding r–subcomposition, where $i = 1, 2, \ldots, C$. Then

$$P_C(\mathbf{Z}(\mathbf{u})) = (\mathbf{s}_1(\mathbf{u}), \ldots, \mathbf{s}_C(\mathbf{u}); \mathbf{V}(\mathbf{u})),$$

will denote the r-partition of order C of $\mathbf{Z}(\mathbf{u})$.

See Aitchison (1986) for a more detailed discussion of additional properties of amalgamations, partitions, subcompositions and other aspects related to compositions that can be extended directly to r-compositions.

An r-amalgamation represents a description, using mathematical symbols, of that which is understood in the figurative sense as an amalgamation: the mixture and combination of several components. Correspondingly, an r-partition is merely a description of the process itself. First the measurement vector is separated into subvectors, then the components of the subvector are added: an amalgamation is the result. At the same time, the information contained in the subvectors is not lost, but rather it is preserved for the statistical analysis. It is summarized in the form of r-subcompositions and the r-composition formed by the sums.

2.5 alr and clr transformations

In the next section log-stationary and log-intrinsic random functions, as well as additive logratio stationary and additive logratio intrinsic vector random functions, are addressed. For this, the alr transformation as defined by Aitchison (1986, p. 113) is required. The clr transformation is also used later, and is therefore referenced at this point.

Definition 2.10 *The* additive logratio (alr) *transformation is defined as:*

$$\begin{aligned} \text{alr} : \mathcal{S}^D &\longrightarrow \mathbb{R}^{D-1} \\ \mathbf{Z}(\mathbf{u}) &\longmapsto \text{alr}(\mathbf{Z}(\mathbf{u})) = \mathbf{X}(\mathbf{u}) = \mathbf{F}\ln(\mathbf{Z}(\mathbf{u})), \end{aligned}$$

where

$$\mathbf{F} = (\mathbf{I} : -\mathbf{j}) = \begin{pmatrix} 1 & 0 & \cdots & 0 & -1 \\ 0 & 1 & \cdots & 0 & -1 \\ \vdots & \vdots & \ddots & \vdots & \vdots \\ 0 & 0 & \cdots & 1 & -1 \end{pmatrix},$$

and **I** *is the* $(D-1) \times (D-1)$ *identity matrix.*

Note that, due to property 2.2, the alr transformation could be defined as well on the basis as

$$\begin{aligned} \text{alr} : \mathbb{R}_+^D &\longrightarrow \mathbb{R}^{D-1} \\ \mathbf{W}(\mathbf{u}) &\longmapsto \text{alr}(\mathbf{W}(\mathbf{u})) = \mathbf{X}(\mathbf{u}) = \mathbf{F}\ln(\mathbf{W}(\mathbf{u})), \end{aligned}$$

leading to the same result.

For every component, the alr transformation implies dividing by the D-th component, followed by taking the (natural) logarithm:

$$\text{alr}(\mathbf{Z}(\mathbf{u})) = \mathbf{X}(\mathbf{u}) = \left(\ln\left(\frac{Z_1(\mathbf{u})}{Z_D(\mathbf{u})}\right) \quad \cdots \quad \ln\left(\frac{Z_{D-1}(\mathbf{u})}{Z_D(\mathbf{u})}\right)\right)',$$

or

$$\text{alr}(\mathbf{W}(\mathbf{u})) = \left(\ln\left(\frac{W_1(\mathbf{u})}{W_D(\mathbf{u})}\right) \quad \cdots \quad \ln\left(\frac{W_{D-1}(\mathbf{u})}{W_D(\mathbf{u})}\right)\right)'.$$

The vector

$$\mathbf{X}(\mathbf{u}) = (\, X_1(\mathbf{u}) \quad X_2(\mathbf{u}) \quad \cdots \quad X_{D-1}(\mathbf{u})\,)'$$

has one component less than the r-composition $\mathbf{Z}(\mathbf{u})$; therefore the auxiliary vector

$$\mathbf{Y}(\mathbf{u}) = (\, X_1(\mathbf{u}) \quad X_2(\mathbf{u}) \quad \cdots \quad X_{D-1}(\mathbf{u}) \quad 0\,)' \tag{2.2}$$

is introduced to define the inverse or *additive generalized logistic* (agl) transformation. Thus, in vector notation, the inverse transformation can be written:

$$\begin{aligned} \text{agl} : \mathbb{R}^{D-1} &\longrightarrow \mathcal{S}^D \\ \mathbf{X}(\mathbf{u}) &\longmapsto \mathbf{Z}(\mathbf{u}) = \frac{\exp(\mathbf{Y}(\mathbf{u}))}{\mathbf{j}'\exp(\mathbf{Y}(\mathbf{u}))}, \end{aligned} \tag{2.3}$$

where

$$\exp(\mathbf{Y}(\mathbf{u})) = (\, \exp(X_1(\mathbf{u})) \quad \exp(X_2(\mathbf{u})) \quad \cdots \quad \exp(X_{D-1}(\mathbf{u})) \quad 1\,)'.$$

If the constant-sum contraint of the r-composition $\mathbf{Z}(\mathbf{u})$ is $c \neq 0$, then the back transformation $\mathbf{Z}(\mathbf{u}) = c \cdot \text{agl}(\mathbf{X}(\mathbf{u}))$ must be used.

Definition 2.11 *The* centered logratio (clr) *transformation is defined as:*

$$\begin{aligned} \text{clr} : \mathbb{R}_+^D &\longrightarrow \mathbb{R}^D \\ \mathbf{W}(\mathbf{u}) &\longmapsto \text{clr}(\mathbf{W}(\mathbf{u})) = \ln\left(\frac{\mathbf{W}(\mathbf{u})}{g(\mathbf{W}(\mathbf{u}))}\right) = \mathbf{G}\ln(\mathbf{W}(\mathbf{u})), \end{aligned}$$

or

$$\begin{aligned} \text{clr} : \mathcal{S}^D &\longrightarrow \mathbb{R}^D \\ \mathbf{Z}(\mathbf{u}) &\longmapsto \text{clr}(\mathbf{Z}(\mathbf{u})) = \ln\left(\frac{\mathbf{Z}(\mathbf{u})}{g(\mathbf{Z}(\mathbf{u}))}\right) = \mathbf{G}\ln(\mathbf{Z}(\mathbf{u})), \end{aligned}$$

where $g(\mathbf{W}(\mathbf{u})) = (\prod_{i=1}^D W_i(\mathbf{u}))^{1/D}$ *is the geometric mean of the components of the basis* $\mathbf{W}(\mathbf{u})$*,* $g(\mathbf{Z}(\mathbf{u})) = (\prod_{i=1}^D Z_i(\mathbf{u}))^{1/D}$ *is the geometric mean of the components of the r-composition* $\mathbf{Z}(\mathbf{u})$ *and*

$$\mathbf{G} = (\mathbf{I} - D^{-1}\mathbf{J}) = \begin{pmatrix} 1-1/D & -1/D & \cdots & -1/D \\ -1/D & 1-1/D & \cdots & -1/D \\ \vdots & \vdots & \ddots & \vdots \\ -1/D & -1/D & \cdots & 1-1/D \end{pmatrix},$$

with $\mathbf{J}$ *the* $D \times D$ *matrix of unities.*

2.6 Hypothesis of stationarity

Consider an arbitrary vector random function $\mathbf{W}(\mathbf{u})$, $\mathbf{u} \in \Omega \subset \mathbb{R}^n$ and a function f of $\mathbf{W}(\mathbf{u})$, such that $f(\mathbf{W}(\mathbf{u}))$ is a vector random function; e.g., if the sample space of $\mathbf{W}(\mathbf{u})$ is $\mathbb{R}_+^D$, f might be the natural logarithm defined as

$$\ln(\mathbf{W}(\mathbf{u})) = (\ln(W_1(\mathbf{u})) \quad \ln(W_2(\mathbf{u})) \quad \cdots \quad \ln(W_D(\mathbf{u})))',$$

and then $f(\mathbf{W}(\mathbf{u}))$ is a D-dimensional vector random function in $\mathbb{R}^D$, or, if the sample space of $\mathbf{W}(\mathbf{u})$ is $\mathcal{S}^D$, f might be the alr transformation defined in the previous section, and then $f(\mathbf{W}(\mathbf{u}))$ reduces to a $(D-1)$-dimensional vector random function in $\mathbb{R}^{D-1}$. The hypothesis of stationarity for $\mathbf{W}(\mathbf{u})$ assumes that the probability distribution of $\mathbf{W}(\mathbf{u})$ is independent of its location $\mathbf{u}$, and requires by definition both the stationarity for each of its components and the joint stationarity, or pairwise stationary correlation (Yaglom 1986, p. 310), of all its components. Recall that knowledge of the marginal cumulative distribution functions of a random vector does not imply knowledge of the joint cumulative distribution function, and therefore stationarity of each component is not enough. In the following when we refer to the stationarity of a vector random function it will be understood that both are included. Here only second-order stationarity (also known as *weak* or *wide-sense* stationarity) and the intrinsic hypothesis are considered, which is all that is required for the oncoming formulations and inference of moments.

Definition 2.12 $\mathbf{W}(\mathbf{u})$ *is* second-order f-stationary *if* $f(\mathbf{W}(\mathbf{u}))$ *is second-order stationary,* i.e., *if* $f(\mathbf{W}(\mathbf{u}))$ *satisfies the following conditions:*

1. *the vector of expected values,* $\mathrm{E}[f(\mathbf{W}(\mathbf{u}))] = \vec{\mu}$, *exists and does not depend on* $\mathbf{u}$;

2. *a matrix-valued covariance function* $\mathrm{Cov}[f(\mathbf{W}(\mathbf{u}_1)), f(\mathbf{W}(\mathbf{u}_2))] = \mathbf{\Sigma}(\mathbf{u}_2 - \mathbf{u}_1)$ *exists and does not depend on* $\mathbf{u}_1$, $\mathbf{u}_2$ *but only on* $\mathbf{h} = \mathbf{u}_2 - \mathbf{u}_1$.

Definition 2.13 $\mathbf{W}(\mathbf{u})$ *is f-intrinsic if* $f(\mathbf{W}(\mathbf{u}))$ *is intrinsic,* i.e., *if* $f(\mathbf{W}(\mathbf{u}))$ *satisfies the following conditions:*

1. *the vector of expected values,* $\mathrm{E}[f(\mathbf{W}(\mathbf{u}))] = \vec{\mu}$, *exists and does not depend on* $\mathbf{u}$;

2. *the matrix-valued covariance function,*

$$\mathrm{Cov}[f(\mathbf{W}(\mathbf{u}_2)) - f(\mathbf{W}(\mathbf{u}_1))] = \mathbf{\Gamma}(\mathbf{u}_2 - \mathbf{u}_1),$$

exists and does not depend on $\mathbf{u}_1$, $\mathbf{u}_2$ *but only on* $\mathbf{h} = \mathbf{u}_2 - \mathbf{u}_1$.

$\boldsymbol{\Sigma}(\mathbf{u}_2 - \mathbf{u}_1)$ is the matrix of cross-covariance functions and $\boldsymbol{\Gamma}(\mathbf{u}_2 - \mathbf{u}_1)$ is the matrix of semivariograms and cross-semivariograms between the components of $f(\mathbf{W}(\mathbf{u}_1))$ and $f(\mathbf{W}(\mathbf{u}_2))$, $\mathbf{u}_1, \mathbf{u}_2 \in \Omega$.

In the following, $\mathbf{W}(\mathbf{u})$ denotes an r-basis and $\mathbf{Z}(\mathbf{u})$ the r-composition that goes with it. Furthermore, logarithmic transformation refers always to the transformation based on the natural logarithm.

Property 2.8 *If* $\mathbf{W}(\mathbf{u})$ *is second-order log-stationary, then the set of all pairwise logratios,*

$$\left\{\ln\left(\frac{W_i(\mathbf{u})}{W_j(\mathbf{u})}\right); i, j = 1, 2, \ldots, D\right\},$$

is second-order stationary.

Proof. To establish second-order stationarity of the logratios, the two properties stated in definition 2.12 have to be verified. For the first property we have

$$\mathrm{E}\left[\ln\left(\frac{W_i(\mathbf{u})}{W_j(\mathbf{u})}\right)\right] = \mathrm{E}[\ln(W_i(\mathbf{u})) - \ln(W_j(\mathbf{u}))] = \mu_i - \mu_j,$$

and the second property can be expressed as

$$\mathrm{Cov}\left[\ln\left(\frac{W_i(\mathbf{u})}{W_k(\mathbf{u})}\right), \ln\left(\frac{W_j(\mathbf{u}+\mathbf{h})}{W_\ell(\mathbf{u}+\mathbf{h})}\right)\right] = \omega_{ij}(\mathbf{h}) - \omega_{i\ell}(\mathbf{h}) - \omega_{kj}(\mathbf{h}) + \omega_{k\ell}(\mathbf{h}),$$

where $\omega_{ij}(\mathbf{h}) = \mathrm{Cov}[\ln(W_i(\mathbf{u})), \ln(W_j(\mathbf{u}+\mathbf{h}))]$. Consequently, the expected value exists and does not depend on $\mathbf{u}$, and the covariance exists and depends only on $\mathbf{h}$, thus completing the proof. □

Definition 2.14 $\mathbf{W}(\mathbf{u})$ *is second-order logratio stationary (*lr *stationary), respectively logratio intrinsic (*lr *intrinsic), if the set of all pairwise logratios introduced in property 2.8 is second-order stationary, respectively intrinsic.*

Property 2.9 *If* $\mathbf{W}(\mathbf{u})$ *is second-order log-stationary (log-intrinsic), then* $\mathbf{W}(\mathbf{u})$ *is second-order* alr *stationary (*alr *intrinsic).*

Proof. The proof of the second-order stationary case is analogous to that for property 2.8, noting that in the first part equalities hold in particular for $j = D$, in the second part they hold for $k = \ell = D$, and for the intrinsic case by applying the same steps to the definition of alr intrinsic and log-intrinsic. □

Property 2.10 $\mathbf{W}(\mathbf{u})$ *is second-order* lr *stationary if and only if* $\mathbf{W}(\mathbf{u})$ *is second-order* alr *stationary.*

Proof. If $\mathbf{W}(\mathbf{u})$ is second-order lr stationary, then in particular the subset of all the logratios formed by dividing with the D-th component is second-order stationary, and therefore $\mathbf{W}(\mathbf{u})$ is second-order alr stationary. Conversely, if $\mathbf{W}(\mathbf{u})$ is second-order alr stationary then, using the fact that

$$\ln\left(\frac{W_i(\mathbf{u})}{W_j(\mathbf{u})}\right) = \ln\left(\frac{W_i(\mathbf{u})}{W_D(\mathbf{u})}\right) - \ln\left(\frac{W_j(\mathbf{u})}{W_D(\mathbf{u})}\right)$$

and substituting it into the definition, it is easily seen that second-order alr stationarity implies second-order lr stationarity. □

Note that the above statements hold for r-bases as well as for r-compositions, and combining them with property 2.2, the proof of the following property is immediate:

Property 2.11 *The following relations hold for an r-basis* $\mathbf{W}(\mathbf{u})$ *and the r-composition* $\mathbf{Z}(\mathbf{u})$ *that goes with it:*

1. $\mathbf{W}(\mathbf{u})$ *second-order log-stationary* $\Longrightarrow$ $\mathbf{Z}(\mathbf{u})$ *second-order* lr *stationary;*
2. $\mathbf{W}(\mathbf{u})$ *second-order log-stationary* $\Longrightarrow$ $\mathbf{Z}(\mathbf{u})$ *second-order* alr *stationary;*
3. $\mathbf{W}(\mathbf{u})$ *second-order* lr *stationary* $\Longleftrightarrow$ $\mathbf{Z}(\mathbf{u})$ *second-order* lr *stationary;*
4. $\mathbf{W}(\mathbf{u})$ *second-order* alr *stationary* $\Longleftrightarrow$ $\mathbf{Z}(\mathbf{u})$ *second-order* alr *stationary.*

The same statements hold for the intrinsic case. Recall that second-order stationarity of a vector random function implies the intrinsic hypothesis for the vector random function. In geostatistics, usually the hypotheses of second-order stationarity and the intrinsic hypothesis are considered, as either of these is sufficient for a spatial-structure analysis and estimation of a coregionalization. The univariate case is just a special case. In practice, it is not possible to test these assumptions; only inconsistencies can be observed. Thus, hypotheses of stationarity of any type are actually decisions or assumptions on the user's side (Deutsch and Journel 1998).

2.7 The additive logistic normal distribution

Most statistical procedures applied in geology are based on the uni- or multivariate normal distribution. As explained in Chapter 5, the cokriging estimator is only optimal if the vector random function has a multivariate normal distribution, where optimal has to be understood in the sense that the estimated values agree with the expected values conditional on the sample. This result was also the case in Journel and Huijbregts (1978), Cressie (1991), and Diggle et al. (1998). For this reason, data corresponding to a clearly nonnormal vector random function are transformed by a suitable transformation in such a way that an underlying multivariate normal distribution can be assumed. The transformation most frequently used is the natural logarithm. The theory of lognormal distributions is found in Aitchison and Brown (1957). The characteristics important for lognormal cokriging are outlined in Chapter 5. The additive logistic normal distribution is introduced here only briefly. The theory can be found in Aitchison (1986, pp. 113–114) and in Aitchison and Shen (1980).

Definition 2.15 *A D-part composition* $\mathbf{Z}$ *has an* additive logistic normal distribution,

$$\mathbf{Z} \sim \mathcal{L}_{D-1}(\vec{\mu}, \mathbf{\Sigma}),$$

if $\mathbf{X} = \mathrm{alr}(\mathbf{Z})$ *has a multivariate normal distribution,* i.e., *if*

$$\mathrm{alr}(\mathbf{Z}) = \mathbf{X} \sim \mathcal{N}_{D-1}(\vec{\mu}, \mathbf{\Sigma}).$$

The density function of $\mathbf{X} \in \mathbb{R}^{D-1}$ is then:

$$(2\pi)^{-d/2}|\mathbf{\Sigma}|^{-1/2}\exp(-\tfrac{1}{2}(\mathbf{X}-\vec{\mu})'\mathbf{\Sigma}^{-1}(\mathbf{X}-\vec{\mu})).$$

From this the density function for $\mathbf{Z} \in \mathcal{S}^D$ follows:

$$(2\pi)^{-d/2}|\mathbf{\Sigma}|^{-1/2}(Z_1\cdots Z_D)^{-1}\exp(-\tfrac{1}{2}(\mathrm{alr}(\mathbf{Z})-\vec{\mu})'\mathbf{\Sigma}^{-1}(\mathrm{alr}(\mathbf{Z})-\vec{\mu})).$$

All moments and all logarithmic moments exist

$$\mathrm{E}\left[\prod_{i=1}^{D} Z_i^{\alpha_i}\right] \quad \text{and} \quad \mathrm{E}\left[\prod_{i=1}^{D} (\ln(Z_i))^{\alpha_i}\right],$$

for $\alpha_i > 0$, $i = 1, \ldots, D$. The integrals, however, are not in a form that is simple to evaluate.

If $\mathbf{Z}$ is additive logistic normally distributed, it follows from the definition that (Z_i/Z_D) and (Z_i/Z_j) are lognormally distributed. So, if interest lies in the ratios, advantage can be taken of the theory of lognormal distributions. The covariance parameter $\mathbf{\Sigma}$ of the $\mathcal{L}$-distribution corresponds to the alr cross-covariance matrix $\mathbf{\Sigma}(\mathbf{h})$ of an r-composition for $\mathbf{h} = 0$ (see Section 3.1.4, definition 3.4). For a further discussion of the characteristics of the additive logistic normal distribution, see Aitchison and Shen (1980) and Aitchison (1986).

Chapter 3

Spatial covariance structure

For any component in time series analysis (Natke 1983), the concept of *covariance* between components of a spatially distributed random vector $\mathbf{Z}(\mathbf{u})$ leads to: *direct covariances,* $\text{Cov}\left[Z_i(\mathbf{u}), Z_j(\mathbf{u})\right]$; *shifted covariances* or *spatial covariances,* $\text{Cov}\left[Z_i(\mathbf{u}), Z_j(\mathbf{u}+\mathbf{h})\right]$, also known as *cross-covariance functions;* and *autocovariance functions,* $\text{Cov}\left[Z_i(\mathbf{u}), Z_i(\mathbf{u}+\mathbf{h})\right]$. The direct covariances may be thought of as a special case of the cross-covariance functions (for $\mathbf{h} = 0$), and the same holds for the autocovariance functions (for $i = j$), so there is no need for a separate discussion. To simplify the exposition, hereafter the term *function* is dropped, and only the terms *cross-covariance* and *autocovariance* are used.

Pawlowsky (1984) stated that if the vector random function constitutes an r-composition, then the problem of spurious spatial correlations appears. This is evident from the fact that at each point of the domain Ω, as in the nonregionalized case, the natural sample space of an r-composition is the D-simplex. This aspect will be discussed in Section 3.1.1.

Aitchison (1986) discussed the problematic nature of the covariance analysis of nonregionalized compositions. He circumvents the problem of spurious correlations by using the fact that the ratio of two arbitrary components of a basis is identical to the ratios of the corresponding components of the associated composition. To avoid working with ratios, which is always difficult, Aitchison takes logarithms of the ratios. Then dependencies among variables of a composition can be examined in real space by analyzing the covariance structure of the log-quotients. The advantages of using this approach are not only numerical or related to the facility of subsequent mathematical operations. Essentially they relate to the fact that the approach consists of a projection of the original sample space, the simplex $\mathcal{S}^D$, onto a new sample space, namely real space $\mathbb{R}^{D-1}$. Thus the door is open to many available methods and models based on the multivariate normal distribution. Recall that the multivariate normal distribution requires the sample space to be precisely the multidimensional, unconstrained real space. For this kind of model, strictly speaking, this is equivalent to saying that you need unconstrained components of the random vector to be analyzed.

One might argue that methods based on the normal distribution are robust; nevertheless, it is important to be aware that robust means that *not-too-strong* departures from the assumed model are allowed. How can this aspect be checked? Applying a multivariate normal model to constrained data is equivalent to using a truncated model formally where the truncation covers an area with probability of occurrence near zero. Thus it would be advisable to check whether truncation is significant before proceeding. Analogous considerations can be used when dealing with one-sided constrained distributions (e.g., the lognormal or gamma distribution) and data that exhibit two-sided constraints (compositional data).

These observations retain their full validity for r-compositions. That is why Aitchison's theory for the statistical analysis of compositions can be extended, making possible the analysis of the spatial covariance structure of r-compositions.

Besides the possibility of an asymmetric spatial covariance structure—which does not occur with direct covariances—the first problem (Section 3.1) is the many potential functions needed to represent the spatial covariance structure, namely D^4 with D variables.

Special specifications for the *spatial covariance structure* derived from the nonregionalized case lead, in a natural manner, to a reduction of the total number of distinct functions that have to be determined. One possible specification which uses lr autocovariances proves to be unsuitable in general because, by definition, it hides any potential asymmetry of the spatial covariance structure. Symmetry is a prerequisite for cokriging in its formulation in terms of semivariograms and cross-semivariograms (Myers 1982). The approach of Yao and Journel (1998) removes this constraint, allowing use of a much broader spectrum of possible models.

The second specification uses alr cross-covariances, retains asymmetry (if present) and also determines the spatial covariance structure uniquely by means of its $(D-1)^2$ functions, which in the case of symmetry reduce to $(D-1)D/2$. Arguing against its use is the fact that the components of the r-composition are considered in an asymmetrical manner; that is, the alr cross-covariances depend on the variable that is chosen as the divisor. The third specification, which uses the clr cross-covariances, is symmetric in the components. It has the same mentioned properties as the second specification, but is not convenient for the estimation procedure being sought because the associated matrix function is singular. Available software neither includes generalized inverses as an option for cokriging nor, more importantly, includes this restriction when searching for admissible models or autocovariance and cross-covariance lookup tables such as those presented by Yao and Journel (1998). Hence, the specification of the spatial covariance structure that uses alr cross-covariances proves to be the most useful. Asymmetry in the components is not a grave disadvantage, as the cokriging procedure is invariant under permutation of the components (see Section 5.4.4).

The intrinsic case must be treated separately because the semivariograms and cross-semivariograms of the logratio transformed variables prevent any possible asymmetry of the spatial structure from being recognized, and are thus, in a first phase, not suitable for spatial-structure analysis. Only after it can be established, at least locally, that nothing contradicts the assumption of symmetry should semivariograms be used for purposes of estimation and modeling.

The possible specifications of the *intrinsic spatial covariance structure* at first glance reveal no clear answer to the question of which is most meaningful for our objectives, and in particular for cokriging. The clr cross-semivariograms should be excluded due to the singularity of the associated matrix function because available software, as mentioned before, does not include generalized inverses as an option. The fact that they are the cross-semivariograms of the alr transformed variables argues for use of the alr cross-semivariograms. On the other hand, lr semivariograms can be estimated and modeled with existing programming techniques. Finally, a further reason supports the choice of this last specification: it opens, in the particular case of r-compositions, new ways of modeling alr cross-covariances, at least for three observed variables which form a composition. This is due to the fact that both pairwise sums and pairwise differences can be expressed as products of log-quotients. This subject is addressed in Section 6.2, after examination of cross associations between the general and the intrinsic spatial covariance structures.

The presentation follows closely the analysis of specifications for the covariance structure of nonregionalized compositions given by Aitchison (1986, chapter 4). Here again proofs are omitted, in general, whenever they are analogous to those of the nonregionalized case or can be directly derived from the corresponding definitions.

From now on "for every lag $\mathbf{h}$" means "for every lag $\mathbf{h}$ such that both ending points $\mathbf{u}_1$ and $\mathbf{u}_2$ of the directional vector $\mathbf{h} = \mathbf{u}_2 - \mathbf{u}_1$ lie in the spatial domain Ω of the r-composition $\mathbf{Z}(\mathbf{u})$." If the explicit reference to lag $\mathbf{h}$ is omitted, it is understood that the statement is made "for every lag **h**." When analyzing symmetry of the spatial covariance structure, the vector **h** might be restricted to a certain range of directions and/or lengths. To simplify exposition, statements in that case will be made for $\mathbf{h} \in \bar{H}$, where $\bar{H}$ stands for the set of considered vectors; e.g., $\bar{H}$ may contain all the admissible vectors in a single direction.

3.1 Second-order stationary case

Throughout this section, consider a D-part second-order lr stationary r–composition $\mathbf{Z}(\mathbf{u})$, $\mathbf{u} \in \Omega$, i.e., the expected values of the pairwise logratios at each point $\mathbf{u}$ exist and do not depend on $\mathbf{u}$, and the cross-covariances between every pairwise logratio at two different points $\mathbf{u}_1$ and $\mathbf{u}_2$ exist and depend only on the directional vector $\mathbf{h} = \mathbf{u}_2 - \mathbf{u}_1$ (see definition 2.14). Recall that by property 2.10, second-order lr stationarity is equivalent to second-order alr stationarity.

3.1.1 Spurious spatial correlation

The main reason to undertake a different approach to spatial correlation or structural analysis of r-compositions is the concept of spurious spatial correlation defined by Pawlowsky (1984). Spurious spatial correlation arises as a consequence of the constant-sum constraint, which is the main characteristic of compositional data in general, and r-compositions in particular. It is expressed as

$$\sum_{i=1}^{D} Z_i(\mathbf{u}) = c, \tag{3.1}$$

as stated in definition 2.1, although there the constant c was assumed equal to one, which is not a restriction to the subsequent approach. From equation 3.1 the singularity of the covariance matrix for any lag **h** is easily derived. First of all, note that the sum of the arithmetic means also satisfies the constant-sum constraint, that is,

$$\sum_{i=1}^{D} \bar{Z}_i(\mathbf{u}) = c. \tag{3.2}$$

Subtracting equations (3.1) and (3.2) we obtain

$$\sum_{i=1}^{D} Z_i(\mathbf{u}) - \sum_{i=1}^{D} \bar{Z}_i(\mathbf{u}) = \sum_{i=1}^{D} (Z_i(\mathbf{u}) - \bar{Z}_i(\mathbf{u})) = 0, \tag{3.3}$$

Multiplying equation (3.3) by $(Z_j(\mathbf{u}+\mathbf{h}) - \bar{Z}_j(\mathbf{u}+\mathbf{h}))$ results in

$$\sum_{i=1}^{D}(Z_i(\mathbf{u}) - \bar{Z}_i(\mathbf{u}))(Z_j(\mathbf{u}+\mathbf{h}) - \bar{Z}_j(\mathbf{u}+\mathbf{h})) = 0, \tag{3.4}$$

and taking expectations on both sides leads to

$$\sum_{i=1}^{D} \mathrm{Cov}(Z_i(\mathbf{u}), Z_j(\mathbf{u}+\mathbf{h})) = 0 \qquad \text{for all } j. \tag{3.5}$$

Equation (3.5) shows that the covariance matrix function is singular for any lag $\mathbf{h}$. We also can see that the negative bias condition—known from the nonregionalized case—appears here: for $i = j$ we have the autocovariance function, which is positive by definition, and the sum being zero means that at least one cross-covariance must be negative. These numerically induced cross-covariances have been termed spurious spatial correlations by Pawlowsky (1984) and should be interpreted accordingly. They are the reason for the search for an alternative approach based on logratios, as developed in following sections.

Note that for a specific pair of components, $Z_i(\mathbf{u})$ and $Z_j(\mathbf{u})$, the cross-covariance function remains the same, whether the whole r-composition or only a subvector is considered. Thus, spurious correlations appear whenever parts of some whole are analyzed, even if they represent only a small fraction of the whole composition. See Chapter 4 for additional comments.

3.1.2 Defining spatial covariance structure

Definition 3.1 *The* spatial covariance structure *of an r-composition* $\mathbf{Z}(\mathbf{u})$ *is the set of* D^4 *functions*

$$\sigma_{ij\cdot k\ell}(\mathbf{h}) = \mathrm{Cov}\left[\ln\left(\frac{Z_i(\mathbf{u})}{Z_k(\mathbf{u})}\right), \ln\left(\frac{Z_j(\mathbf{u}+\mathbf{h})}{Z_\ell(\mathbf{u}+\mathbf{h})}\right)\right], \quad i,j,k,\ell \in \{1,2,\ldots,D\},$$

for $\mathbf{u}, \mathbf{u}+\mathbf{h} \in \Omega$.

Property 3.1 *The spatial covariance structure of* $\mathbf{Z}(\mathbf{u})$ *has the following properties* ($i,j,k,\ell, m,n \in \{1,2,\ldots,D\}$):

1. *in general,* $\sigma_{ij\cdot k\ell}(\mathbf{h}) \neq \sigma_{ji\cdot \ell k}(\mathbf{h})$,

2. *in general,* $\sigma_{ij\cdot k\ell}(\mathbf{h}) \neq \sigma_{ij\cdot k\ell}(-\mathbf{h})$,

3. $\sigma_{ij\cdot k\ell}(\mathbf{h}) = \sigma_{ji\cdot \ell k}(-\mathbf{h})$,

4. $\sigma_{ij\cdot k\ell}(\mathbf{h}) = -\sigma_{kj\cdot i\ell}(\mathbf{h}) = \sigma_{k\ell\cdot ij}(\mathbf{h}) = -\sigma_{i\ell\cdot kj}(\mathbf{h})$,

5. $\sigma_{ij\cdot i\ell}(\mathbf{h}) = \sigma_{ij\cdot kj}(\mathbf{h}) = \sigma_{ij\cdot ij}(\mathbf{h}) = \sigma_{ii\cdot ii}(\mathbf{h}) = 0$,

6. $\sigma_{ij\cdot k\ell}(\mathbf{h}) = \sigma_{ij\cdot mn}(\mathbf{h}) + \sigma_{in\cdot m\ell}(\mathbf{h}) + \sigma_{mj\cdot kn}(\mathbf{h}) + \sigma_{mn\cdot k\ell}(\mathbf{h})$.

Proof. By definition 3.1

$$\sigma_{ij\cdot k\ell}(\mathbf{h}) = \mathrm{Cov}\left[\ln\left(\frac{Z_i(\mathbf{u})}{Z_k(\mathbf{u})}\right), \ln\left(\frac{Z_j(\mathbf{u}+\mathbf{h})}{Z_\ell(\mathbf{u}+\mathbf{h})}\right)\right],$$

and

$$\sigma_{ji\cdot \ell k}(\mathbf{h}) = \mathrm{Cov}\left[\ln\left(\frac{Z_j(\mathbf{u})}{Z_\ell(\mathbf{u})}\right), \ln\left(\frac{Z_i(\mathbf{u}+\mathbf{h})}{Z_k(\mathbf{u}+\mathbf{h})}\right)\right],$$

where $\ln(Z_i(\mathbf{u})/Z_k(\mathbf{u}))$ is a variable at location $\mathbf{u}$ and $\ln(Z_i(\mathbf{u}+\mathbf{h})/Z_k(\mathbf{u}+\mathbf{h}))$ is a variable at location $\mathbf{u}+\mathbf{h}$. They are, therefore, different random variables, independent of the fact that second-order lr stationarity requires them to be identically distributed. The same holds for $\ln(Z_j(\mathbf{u})/Z_\ell(\mathbf{u}))$, which is a variable at location $\mathbf{u}$, while $\ln(Z_j(\mathbf{u}+\mathbf{h})/Z_\ell(\mathbf{u}+\mathbf{h}))$ is a variable at location $\mathbf{u}+\mathbf{h}$, thus these are again different random variables. Consequently, in the two stated cross-covariances, the four variables involved are different, hence there is no reason for them to be equal, leading us to property 3.1.1.

The same arguments lead us to the conclusion that

$$\sigma_{ij\cdot k\ell}(\mathbf{h}) = \mathrm{Cov}\left[\ln\left(\frac{Z_i(\mathbf{u})}{Z_k(\mathbf{u})}\right), \ln\left(\frac{Z_j(\mathbf{u}+\mathbf{h})}{Z_\ell(\mathbf{u}+\mathbf{h})}\right)\right]$$

and

$$\sigma_{ij\cdot k\ell}(-\mathbf{h}) = \mathrm{Cov}\left[\ln\left(\frac{Z_i(\mathbf{u})}{Z_k(\mathbf{u})}\right), \ln\left(\frac{Z_j(\mathbf{u}-\mathbf{h})}{Z_\ell(\mathbf{u}-\mathbf{h})}\right)\right]$$

cannot be identical in general, as stated in property 3.1.2. There exist cases where the identity holds, so we introduce the concept of a symmetric covariance structure, as stated in definition 3.2.

This fact may be illustrated with the aid of diagrams, as shown in Figure 3.1, where each arrow represents a spatial covariance of the type $\sigma_{ij\cdot k\ell}(\mathbf{h})$. It should be noted that the first diagram expresses asymmetry in the indices and the second expresses asymmetry in the sign of the direction vector $\mathbf{h}$. They are mutually dependent, as can be seen from property 3.1.3, for which proof goes as follows.

From definition 3.1 we have

$$\sigma_{ij\cdot k\ell}(\mathbf{h}) = \mathrm{Cov}\left[\ln\left(\frac{Z_i(\mathbf{u})}{Z_k(\mathbf{u})}\right), \ln\left(\frac{Z_j(\mathbf{u}+\mathbf{h})}{Z_\ell(\mathbf{u}+\mathbf{h})}\right)\right],$$

and

$$\sigma_{ji\cdot \ell k}(-\mathbf{h}) = \mathrm{Cov}\left[\ln\left(\frac{Z_j(\mathbf{u})}{Z_\ell(\mathbf{u})}\right), \ln\left(\frac{Z_i(\mathbf{u}-\mathbf{h})}{Z_k(\mathbf{u}-\mathbf{h})}\right)\right]. \tag{3.6}$$

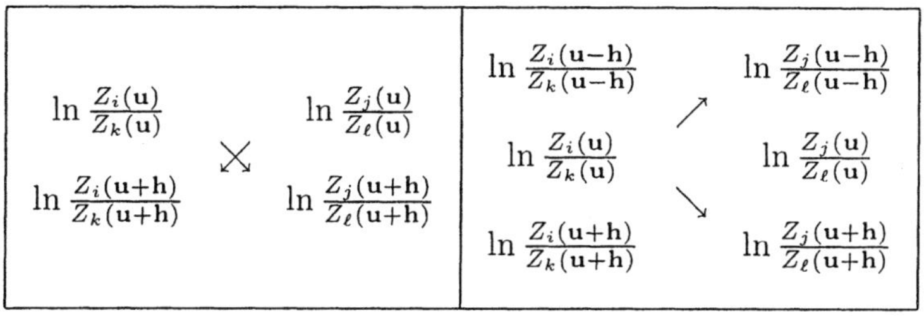

Figure 3.1: Representation of spatial covariances. Left: SE arrow = $\sigma_{ij\cdot k\ell}(\mathbf{h})$; SW arrow = $\sigma_{ji\cdot \ell k}(\mathbf{h})$. Right: NE arrow = $\sigma_{ij\cdot k\ell}(-\mathbf{h})$; SE arrow = $\sigma_{ij\cdot k\ell}(\mathbf{h})$.

As by the hypothesis of second-order lr stationarity, the covariance depends only on the lag $\mathbf{h}$ and not on the spatial location $\mathbf{u}$, let us write in equation (3.6) $\mathbf{u}-\mathbf{h}=\mathbf{u}'$, which implies $\mathbf{u}=\mathbf{u}'+\mathbf{h}$, resulting in

$$\begin{aligned}\sigma_{ji\cdot \ell k}(-\mathbf{h}) &= \mathrm{Cov}\left[\ln\left(\frac{Z_j(\mathbf{u}'+\mathbf{h})}{Z_\ell(\mathbf{u}'+\mathbf{h})}\right), \ln\left(\frac{Z_i(\mathbf{u}')}{Z_k(\mathbf{u}')}\right)\right] \\ &= \mathrm{Cov}\left[\ln\left(\frac{Z_i(\mathbf{u}')}{Z_k(\mathbf{u}')}\right), \ln\left(\frac{Z_j(\mathbf{u}'+\mathbf{h})}{Z_\ell(\mathbf{u}'+\mathbf{h})}\right)\right] \\ &= \sigma_{ij\cdot k\ell}(\mathbf{h}),\end{aligned}$$

after applying the fact that for any pair of random variables X and X', $\mathrm{Cov}\,[X, X'] = \mathrm{Cov}\,[X', X]$.

To state property 3.1.4, we use the fact that for any two random variables X and X' the following equalities hold:

$$\ln\left(\frac{X}{X'}\right) = \ln(X) - \ln(X') = -(\ln(X') - \ln(X)) = -\ln\left(\frac{X'}{X}\right);$$

$$\mathrm{Cov}\,[-X, X'] = \mathrm{Cov}\,[X, -X'] = -\mathrm{Cov}\,[X, X'].$$

Thus, the following sequence of equalities is derived:

$$\begin{aligned}\sigma_{ij\cdot k\ell}(\mathbf{h}) &= \mathrm{Cov}\left[\ln\left(\frac{Z_i(\mathbf{u})}{Z_k(\mathbf{u})}\right), \ln\left(\frac{Z_j(\mathbf{u}+\mathbf{h})}{Z_\ell(\mathbf{u}+\mathbf{h})}\right)\right] \\ &= \mathrm{Cov}\left[-\ln\left(\frac{Z_k(\mathbf{u})}{Z_i(\mathbf{u})}\right), \ln\left(\frac{Z_j(\mathbf{u}+\mathbf{h})}{Z_\ell(\mathbf{u}+\mathbf{h})}\right)\right] \\ &= -\mathrm{Cov}\left[\ln\left(\frac{Z_k(\mathbf{u})}{Z_i(\mathbf{u})}\right), \ln\left(\frac{Z_j(\mathbf{u}+\mathbf{h})}{Z_\ell(\mathbf{u}+\mathbf{h})}\right)\right] \\ &= -\sigma_{kj\cdot i\ell}(\mathbf{h}),\end{aligned}$$

which proves the first equality. Applying the same reasoning twice, the second and third equalities are obtained.

Property 3.1.5 is straightforward, taking into account that for any random variable X, $\ln(X/X) = \ln(1) = 0$ and that the covariance of any random variable with one that is constant is zero.

Finally, property 3.1.6, which is necessary to prove other properties in subsequent steps, is obtained as follows, using again the fact that for any random variable X, $\ln(X/X) = \ln(1) = 0$ and basic properties of the logarithmic and covariance functions:

$$\sigma_{ij\cdot k\ell}(\mathbf{h}) = \mathrm{Cov}\left[\ln\left(\frac{Z_i(\mathbf{u})}{Z_k(\mathbf{u})}\right), \ln\left(\frac{Z_j(\mathbf{u}+\mathbf{h})}{Z_\ell(\mathbf{u}+\mathbf{h})}\right)\right]$$

$$= \mathrm{Cov}\left[\ln\left(\frac{Z_i(\mathbf{u})}{Z_k(\mathbf{u})}\right) + \ln\left(\frac{Z_m(\mathbf{u})}{Z_m(\mathbf{u})}\right), \ln\left(\frac{Z_j(\mathbf{u}+\mathbf{h})}{Z_\ell(\mathbf{u}+\mathbf{h})}\right) + \ln\left(\frac{Z_n(\mathbf{u}+\mathbf{h})}{Z_n(\mathbf{u}+\mathbf{h})}\right)\right]$$

$$= \mathrm{Cov}\left[\ln\left(\frac{Z_i(\mathbf{u})}{Z_m(\mathbf{u})}\right) + \ln\left(\frac{Z_m(\mathbf{u})}{Z_k(\mathbf{u})}\right), \ln\left(\frac{Z_j(\mathbf{u}+\mathbf{h})}{Z_n(\mathbf{u}+\mathbf{h})}\right) + \ln\left(\frac{Z_n(\mathbf{u}+\mathbf{h})}{Z_\ell(\mathbf{u}+\mathbf{h})}\right)\right]$$

$$= \mathrm{Cov}\left[\ln\left(\frac{Z_i(\mathbf{u})}{Z_m(\mathbf{u})}\right), \ln\left(\frac{Z_j(\mathbf{u}+\mathbf{h})}{Z_n(\mathbf{u}+\mathbf{h})}\right)\right] + \mathrm{Cov}\left[\ln\left(\frac{Z_i(\mathbf{u})}{Z_m(\mathbf{u})}\right), \ln\left(\frac{Z_n(\mathbf{u}+\mathbf{h})}{Z_\ell(\mathbf{u}+\mathbf{h})}\right)\right]$$

$$+ \mathrm{Cov}\left[\ln\left(\frac{Z_m(\mathbf{u})}{Z_k(\mathbf{u})}\right), \ln\left(\frac{Z_j(\mathbf{u}+\mathbf{h})}{Z_n(\mathbf{u}+\mathbf{h})}\right)\right] + \mathrm{Cov}\left[\ln\left(\frac{Z_m(\mathbf{u})}{Z_k(\mathbf{u})}\right), \ln\left(\frac{Z_n(\mathbf{u}+\mathbf{h})}{Z_\ell(\mathbf{u}+\mathbf{h})}\right)\right]$$

$$= \sigma_{ij\cdot mn}(\mathbf{h}) + \sigma_{in\cdot m\ell}(\mathbf{h}) + \sigma_{mj\cdot kn}(\mathbf{h}) + \sigma_{mn\cdot k\ell}(\mathbf{h}),$$

thus completing the proof of property 3.1. □

Definition 3.2 *The spatial covariance structure of* $\mathbf{Z}(\mathbf{u})$ *is* symmetric *over* $\bar{H}$ *if, for all* $i, j, k, \ell \in \{1, 2, \ldots, D\}$ *and for every* $\mathbf{h} \in \bar{H}$,

$$\sigma_{ij\cdot k\ell}(\mathbf{h}) = \sigma_{ji\cdot \ell k}(\mathbf{h}).$$

Property 3.2 *The spatial covariance structure of* $\mathbf{Z}(\mathbf{u})$ *is symmetric over* $\bar{H}$ *if and only if for all* $i, j, k, \ell \in \{1, 2, \ldots, D\}$ *and for every* $\mathbf{h} \in \bar{H}$,

$$\sigma_{ij\cdot k\ell}(\mathbf{h}) = \sigma_{ij\cdot k\ell}(-\mathbf{h}).$$

Proof. Using property 3.1.3, the proof is straightforward in both directions. □

Note that the symmetry of the spatial covariance structure is expressed as a symmetry in the indices, while property 3.2 expresses a symmetry in direction; that is, the symmetry of the spatial covariance structure of $\mathbf{Z}(\mathbf{u})$ does not depend on the sign of the directional vector $\mathbf{h}$, and $\bar{H}$ must also contain $-\mathbf{h}$ with every $\mathbf{h}$. Furthermore, recall that $0 \in \bar{H}$ at all times.

In a manner completely parallel to the nonregionalized case, definition 3.1 yields D^4 different functions $\sigma_{ij\cdot k\ell}(\mathbf{h})$ corresponding to possible combinations of the indices i, j, k, ℓ. In this instance, the properties of the spatial covariance structure listed in definition 3.1 lead to a large decrease in total number of auto- and cross-covariances that are not functionally interdependent. However, due to the asymmetry of the spatial covariance structure expressed by properties 3.1.1 and 3.1.2, this decrease is less distinct in the regionalized case.

3.1.3 lr autocovariance

Definition 3.3 *The* lr autocovariance *of any two components $Z_i(\mathbf{u})$, $Z_j(\mathbf{u})$, $i, j = 1, 2, \ldots, D$, of an r-composition $\mathbf{Z}(\mathbf{u})$ is the autocovariance of the corresponding logratio,*

$$\tau_{i\cdot j}(\mathbf{h}) = \sigma_{ii\cdot jj}(\mathbf{h}),$$

and the variation matrix, $\mathbf{T}(\mathbf{h})$, *of* $\mathbf{Z}(\mathbf{u})$ *is the matrix of* lr *autocovariances, with dimension $D \times D$,*

$$\mathbf{T}(\mathbf{h}) = (\,\tau_{i\cdot j}(\mathbf{h})\,)\,.$$

Property 3.3 *The* lr *autocovariances have the following properties for i, j, k, ℓ all different:*

1. $\tau_{i\cdot i}(\mathbf{h}) = 0$,
2. $\tau_{i\cdot j}(\mathbf{h}) = \tau_{j\cdot i}(\mathbf{h}) = \tau_{i\cdot j}(-\mathbf{h})$,
3. $\sigma_{ij\cdot k\ell}(\mathbf{h}) + \sigma_{ji\cdot \ell k}(\mathbf{h}) = \tau_{i\cdot \ell}(\mathbf{h}) + \tau_{j\cdot k}(\mathbf{h}) - \tau_{i\cdot j}(\mathbf{h}) - \tau_{k\cdot \ell}(\mathbf{h})$;
4. *If the spatial covariance structure of $\mathbf{Z}(\mathbf{u})$ is symmetric, then*

$$\sigma_{ij\cdot k\ell}(\mathbf{h}) = \tfrac{1}{2}\left[\tau_{i\cdot \ell}(\mathbf{h}) + \tau_{j\cdot k}(\mathbf{h}) - \tau_{i\cdot j}(\mathbf{h}) - \tau_{k\cdot \ell}(\mathbf{h})\right].$$

Proof. Property 3.3.1 is obtained from definition 3.3 and property 3.1.5, while property 3.3.2 is obtained from definition 3.3 and property 3.1.4 for the first equality and property 3.1.3 for the second equality.

To obtain property 3.3.3 we apply first property 3.1.6 to each summand of the left-hand side, using for the first function $m = \ell$ and $n = i$, while for the second we use $m = k$ and $n = j$ after rearranging the indices according to the expression required:

$$\begin{aligned}
&\sigma_{ij\cdot k\ell}(\mathbf{h}) + \sigma_{ji\cdot \ell k}(\mathbf{h}) \\
&\quad = \sigma_{ij\cdot \ell i}(\mathbf{h}) + \sigma_{ii\cdot \ell\ell}(\mathbf{h}) + \sigma_{\ell j\cdot ki}(\mathbf{h}) + \sigma_{\ell i\cdot k\ell}(\mathbf{h}) \\
&\qquad + \sigma_{ji\cdot kj}(\mathbf{h}) + \sigma_{jj\cdot kk}(\mathbf{h}) + \sigma_{ki\cdot \ell j}(\mathbf{h}) + \sigma_{kj\cdot \ell k}(\mathbf{h})
\end{aligned}$$

$$\begin{aligned} &= -\sigma_{ii\cdot\ell j}(\mathbf{h}) + \tau_{i\cdot\ell}(\mathbf{h}) + \sigma_{\ell j\cdot ki}(\mathbf{h}) + \sigma_{\ell i\cdot k\ell}(\mathbf{h}) \\ &\quad + \sigma_{ji\cdot kj}(\mathbf{h}) + \tau_{j\cdot k}(\mathbf{h}) + \sigma_{ki\cdot\ell j}(\mathbf{h}) - \sigma_{kk\cdot\ell j}(\mathbf{h}) \end{aligned}$$

applying definition 3.3 and property 3.1.4 where appropriate. From property 3.1.6 we have for $m = j$ and $n = j$ and taking into account property 3.1.5:

$$\begin{aligned} -\sigma_{ii\cdot\ell j}(\mathbf{h}) &= -\sigma_{ii\cdot jj}(\mathbf{h}) - \sigma_{ij\cdot jj}(\mathbf{h}) - \sigma_{ji\cdot\ell j}(\mathbf{h}) - \sigma_{jj\cdot\ell j}(\mathbf{h}) \\ &= -\tau_{i\cdot j}(\mathbf{h}) - \sigma_{ji\cdot\ell j}(\mathbf{h}), \end{aligned}$$

and analogously we obtain

$$-\sigma_{kk\cdot\ell j}(\mathbf{h}) = -\tau_{k\cdot\ell}(\mathbf{h}) - \sigma_{k\ell\cdot\ell j}(\mathbf{h}).$$

Thus we have

$$\begin{aligned} &\sigma_{ij\cdot k\ell}(\mathbf{h}) + \sigma_{ji\cdot\ell k}(\mathbf{h}) \\ &= -\tau_{i\cdot j}(\mathbf{h}) - \sigma_{ji\cdot\ell j}(\mathbf{h}) + \tau_{i\cdot\ell}(\mathbf{h}) + \sigma_{\ell j\cdot ki}(\mathbf{h}) + \sigma_{\ell i\cdot k\ell}(\mathbf{h}) \\ &\quad + \sigma_{ji\cdot kj}(\mathbf{h}) + \tau_{j\cdot k}(\mathbf{h}) + \sigma_{ki\cdot\ell j}(\mathbf{h}) - \tau_{k\cdot\ell}(\mathbf{h}) - \sigma_{k\ell\cdot\ell j}(\mathbf{h}) \\ &= \tau_{i\cdot\ell}(\mathbf{h}) + \tau_{j\cdot k}(\mathbf{h}) - \tau_{i\cdot j}(\mathbf{h}) - \tau_{k\cdot\ell}(\mathbf{h}) \\ &\quad -\sigma_{ji\cdot\ell j}(\mathbf{h}) + \sigma_{\ell j\cdot ki}(\mathbf{h}) + \sigma_{\ell i\cdot k\ell}(\mathbf{h}) + \sigma_{ji\cdot kj}(\mathbf{h}) + \sigma_{ki\cdot\ell j}(\mathbf{h}) - \sigma_{k\ell\cdot\ell j}(\mathbf{h}), \end{aligned}$$

and to prove property 3.3.3, we need to see that

$$-\sigma_{ji\cdot\ell j}(\mathbf{h}) + \sigma_{\ell j\cdot ki}(\mathbf{h}) + \sigma_{\ell i\cdot k\ell}(\mathbf{h}) + \sigma_{ji\cdot kj}(\mathbf{h}) + \sigma_{ki\cdot\ell j}(\mathbf{h}) - \sigma_{k\ell\cdot\ell j}(\mathbf{h}) = 0. \quad (3.7)$$

But, using once more property 3.1.6 with $m = k$ and $n = j$ combined with property 3.1.5, we obtain

$$-\sigma_{ji\cdot\ell j}(\mathbf{h}) = -\sigma_{ji\cdot kj}(\mathbf{h}) - \sigma_{ki\cdot\ell j}(\mathbf{h}),$$

and, taking first into account property 3.1.4, we see that

$$-\sigma_{k\ell\cdot\ell j}(\mathbf{h}) = -\sigma_{\ell j\cdot k\ell}(\mathbf{h}) = -\sigma_{\ell j\cdot ki}(\mathbf{h}) - \sigma_{\ell i\cdot k\ell}(\mathbf{h}).$$

Substituting these results into equation (3.7), we see that the required equality holds. The last property is directly obtained using definition 3.2 and dividing both sides by two. □

The symmetry in the indices and in $\mathbf{h}$ of the lr autocovariance for any two components of $\mathbf{Z}(\mathbf{u})$, expressed in property 3.3.2, should be pointed out. Therefore the use of lr autocovariances, or the use of the variation matrix, to describe the spatial behavior of an r-composition before ensuring symmetry would mean an important information loss, because a possible asymmetry of the spatial covariance structure is concealed. This is evident from property 3.3.3, which states that if the spatial covariance structure is not symmetric, then it is not possible to completely specify it through the knowledge of the lr autocovariances. Another specification—which allows recognition of asymmetry, if present—is required to reduce the total number of functions necessary to describe the spatial covariance structure of an r-composition.

3.1.4 alr cross-covariance

Definition 3.4 *For two components* $Z_i(\mathbf{u})$, $Z_j(\mathbf{u})$, $i, j = 1, 2, \ldots, D$, *of an r-composition* $\mathbf{Z}(\mathbf{u})$ *the* alr cross-covariance *is the function*

$$\sigma_{ij}(\mathbf{h}) = \sigma_{ij \cdot DD}(\mathbf{h}).$$

The alr autocovariance *of any component is included as a particular case for* $i = j$.

The alr cross-covariance matrix *of* $\mathbf{Z}(\mathbf{u})$ *is the matrix-valued function, with dimension* $(D-1) \times (D-1)$, *of the* alr *cross-covariances for every possible pair of components,*

$$\boldsymbol{\Sigma}(\mathbf{h}) = (\, \sigma_{ij}(\mathbf{h})\,).$$

Property 3.4 *The* alr *cross-covariances have the following properties:*

1. *in general* $\boldsymbol{\Sigma}(\mathbf{h}) \neq \boldsymbol{\Sigma}'(\mathbf{h})$,
2. *in general* $\boldsymbol{\Sigma}(\mathbf{h}) \neq \boldsymbol{\Sigma}(-\mathbf{h})$,
3. $\boldsymbol{\Sigma}(\mathbf{h}) = \boldsymbol{\Sigma}'(-\mathbf{h})$,
4. $\sigma_{ij \cdot k\ell}(\mathbf{h}) = \sigma_{ij}(\mathbf{h}) + \sigma_{k\ell}(\mathbf{h}) - \sigma_{i\ell}(\mathbf{h}) - \sigma_{kj}(\mathbf{h})$.

Proof. Properties 3.4.1, 3.4.2 and 3.4.3 are identical to properties 3.1.1, 3.1.2 and 3.1.3, but for alr cross-covariances and expressed in matrix notation, while property 3.4.4 is obtained using property 3.1.6 with $m = n = D$ and then property 3.1.4. □

The alr autocovariances $\sigma_{ii}(\mathbf{h})$, $i = 1, 2, \ldots, D-1$, coincide with the lr autocovariances $\tau_{i \cdot D}(\mathbf{h})$ and therefore satisfy property 3.3.2, while the alr cross-covariances of different components evidence asymmetry of the spatial covariance structure, if present; this is so because, as shall be seen later, asymmetry of the spatial covariance structure is equivalent to inequality in properties 3.4.1 and 3.4.2. As indicated in property 3.4.4, the spatial covariance structure also can be uniquely expressed in terms of the alr cross-covariances. Therefore the alr cross-covariances are better suited for analyzing the spatial covariance structure in the regionalized case than are lr autocovariances. If the spatial covariance structure is not symmetric, then all $(D-1)^2$ functions of the alr cross-covariance matrix are required to completely determine the spatial covariance structure of $\mathbf{Z}(\mathbf{u})$. If it is symmetric, then the number reduces to $(D-1)D/2$.

Definition 3.5 *The* alr *cross-covariance of two components* $Z_i(\mathbf{u})$ *and* $Z_j(\mathbf{u})$, $i \neq j$, *of* $\mathbf{Z}(\mathbf{u})$ *is* symmetric *over* $\bar{H}$ *if, for every* $\mathbf{h} \in \bar{H}$,

$$\sigma_{ij}(\mathbf{h}) = \sigma_{ji}(\mathbf{h});$$

accordingly, the alr *cross-covariance matrix is symmetric over* $\bar{H}$ *if, for every* $\mathbf{h} \in \bar{H}$,

$$\mathbf{\Sigma}(\mathbf{h}) = \mathbf{\Sigma}'(\mathbf{h}).$$

This concept of symmetry in the indices of the alr cross-covariances corresponds to the usual concept of symmetry of a matrix for the alr cross-covariance matrix for every lag $\mathbf{h} \in \bar{H}$. Furthermore, as can be seen easily from the definition and property 3.4.3, symmetry in the indices is equivalent to symmetry in direction:

Property 3.5 *For every lag* $\mathbf{h}$,

$$\begin{aligned} \sigma_{ij}(\mathbf{h}) = \sigma_{ji}(\mathbf{h}) \quad &\Longleftrightarrow \quad \sigma_{ij}(\mathbf{h}) = \sigma_{ij}(-\mathbf{h}); \\ \mathbf{\Sigma}(\mathbf{h}) = \mathbf{\Sigma}'(\mathbf{h}) \quad &\Longleftrightarrow \quad \mathbf{\Sigma}(\mathbf{h}) = \mathbf{\Sigma}(-\mathbf{h}). \end{aligned}$$

3.1.5 clr cross-covariance

The fact that alr cross-covariances depend on the component chosen as the divisor leads to the following definition, which is symmetric in the components:

Definition 3.6 *For two components* $Z_i(\mathbf{u})$, $Z_j(\mathbf{u})$, $i, j = 1, 2, \ldots, D$, *of an r-composition* $\mathbf{Z}(\mathbf{u})$ *the* clr cross-covariance *is the function*

$$\xi_{ij}(\mathbf{h}) = \operatorname{Cov}\left[\ln\left(\frac{Z_i(\mathbf{u})}{g(\mathbf{Z}(\mathbf{u}))}\right), \ln\left(\frac{Z_j(\mathbf{u}+\mathbf{h})}{g(\mathbf{Z}(\mathbf{u}+\mathbf{h}))}\right)\right],$$

where $g(\mathbf{Z}(\mathbf{u}))$ *is the geometric mean of the components as introduced in definition 2.11; the* clr autocovariance *of any component is included as a particular case for* $i = j$. *The* clr cross-covariance matrix, *with size* $D \times D$, *is then the matrix of the* clr *cross-covariances for every possible pair of components,*

$$\mathbf{\Xi}(\mathbf{h}) = (\,\xi_{ij}(\mathbf{h})\,).$$

Property 3.6 *The* clr *cross-covariances have the following properties:*

1. *in general,* $\boldsymbol{\Xi}(\mathbf{h}) \neq \boldsymbol{\Xi}'(\mathbf{h})$;
2. *in general,* $\boldsymbol{\Xi}(\mathbf{h}) \neq \boldsymbol{\Xi}(-\mathbf{h})$;
3. $\boldsymbol{\Xi}(\mathbf{h}) = \boldsymbol{\Xi}'(-\mathbf{h})$;
4. $\sigma_{ij\cdot k\ell}(\mathbf{h}) = \xi_{ij}(\mathbf{h}) + \xi_{k\ell}(\mathbf{h}) - \xi_{i\ell}(\mathbf{h}) - \xi_{kj}(\mathbf{h})$;
5. $\boldsymbol{\Xi}(\mathbf{h})\mathbf{j} = \mathbf{0}$ *(the* clr *cross-covariance matrix is singular for all* $\mathbf{h}$*).*

Proof. Arguments identical to those used to prove properties 3.1.1 and 3.1.2 show that properties 3.6.1 and 3.6.2 hold and are not repeated here. It may be seen that property 3.6.3 is true by noting that under the hypothesis of second-order clr stationarity for any indices i, j, $i \neq j$, taking $\mathbf{u}' = \mathbf{u} + \mathbf{h}$,

$$\begin{aligned}
\xi_{ij}(\mathbf{h}) &= \operatorname{Cov}\left[\ln\left(\frac{Z_i(\mathbf{u})}{g(\mathbf{Z}(\mathbf{u}))}\right), \ln\left(\frac{Z_j(\mathbf{u}+\mathbf{h})}{g(\mathbf{Z}(\mathbf{u}+\mathbf{h}))}\right)\right] \\
&= \operatorname{Cov}\left[\ln\left(\frac{Z_i(\mathbf{u}'-\mathbf{h})}{g(\mathbf{Z}(\mathbf{u}'-\mathbf{h}))}\right), \ln\left(\frac{Z_j(\mathbf{u}')}{g(\mathbf{Z}(\mathbf{u}'))}\right)\right] \\
&= \operatorname{Cov}\left[\ln\left(\frac{Z_j(\mathbf{u}')}{g(\mathbf{Z}(\mathbf{u}'))}\right), \ln\left(\frac{Z_i(\mathbf{u}'-\mathbf{h})}{g(\mathbf{Z}(\mathbf{u}'-\mathbf{h}))}\right)\right] \\
&= \xi_{ji}(-\mathbf{h}).
\end{aligned}$$

To prove property 3.6.4 we multiply and divide each component by the geometric mean of the composition at the corresponding position, $\mathbf{u}$ and $\mathbf{u} + \mathbf{h}$ in

$$\sigma_{ij\cdot k\ell}(\mathbf{h}) = \operatorname{Cov}\left[\ln\left(\frac{Z_i(\mathbf{u})}{Z_k(\mathbf{u})}\right), \ln\left(\frac{Z_j(\mathbf{u}+\mathbf{h})}{Z_\ell(\mathbf{u}+\mathbf{h})}\right)\right],$$

and thus obtain the two expressions

$$\begin{aligned}
\ln\left(\frac{Z_i(\mathbf{u})}{Z_k(\mathbf{u})}\right) &= \ln\left(\frac{Z_i(\mathbf{u})g(\mathbf{Z}(\mathbf{u}))}{g(\mathbf{Z}(\mathbf{u}))Z_k(\mathbf{u})}\right) \\
&= \ln\left(\frac{Z_i(\mathbf{u})}{g(\mathbf{Z}(\mathbf{u}))}\right) - \ln\left(\frac{Z_k(\mathbf{u})}{g(\mathbf{Z}(\mathbf{u}))}\right), \\
\ln\left(\frac{Z_j(\mathbf{u}+\mathbf{h})}{Z_\ell(\mathbf{u}+\mathbf{h})}\right) &= \ln\left(\frac{Z_j(\mathbf{u}+\mathbf{h})g(\mathbf{Z}(\mathbf{u}+\mathbf{h}))}{g(\mathbf{Z}(\mathbf{u}+\mathbf{h}))Z_\ell(\mathbf{u}+\mathbf{h})}\right) \\
&= \ln\left(\frac{Z_j(\mathbf{u}+\mathbf{h})}{g(\mathbf{Z}(\mathbf{u}+\mathbf{h}))}\right) - \ln\left(\frac{Z_\ell(\mathbf{u}+\mathbf{h})}{g(\mathbf{Z}(\mathbf{u}+\mathbf{h}))}\right).
\end{aligned}$$

Developing the resulting expression of the covariance, we obtain

$$\begin{aligned}
\sigma_{ij\cdot k\ell}(\mathbf{h}) &= \operatorname{Cov}\left[\ln\left(\frac{Z_i(\mathbf{u})}{g(\mathbf{Z}(\mathbf{u}))}\right), \ln\left(\frac{Z_j(\mathbf{u}+\mathbf{h})}{g(\mathbf{Z}(\mathbf{u}+\mathbf{h}))}\right)\right] \\
&\quad - \operatorname{Cov}\left[\ln\left(\frac{Z_i(\mathbf{u})}{g(\mathbf{Z}(\mathbf{u}))}\right), \ln\left(\frac{Z_\ell(\mathbf{u}+\mathbf{h})}{g(\mathbf{Z}(\mathbf{u}+\mathbf{h}))}\right)\right] \\
&\quad - \operatorname{Cov}\left[\ln\left(\frac{Z_k(\mathbf{u})}{g(\mathbf{Z}(\mathbf{u}))}\right), \ln\left(\frac{Z_j(\mathbf{u}+\mathbf{h})}{g(\mathbf{Z}(\mathbf{u}+\mathbf{h}))}\right)\right] \\
&\quad + \operatorname{Cov}\left[\ln\left(\frac{Z_k(\mathbf{u})}{g(\mathbf{Z}(\mathbf{u}))}\right), \ln\left(\frac{Z_\ell(\mathbf{u}+\mathbf{h})}{g(\mathbf{Z}(\mathbf{u}+\mathbf{h}))}\right)\right] \\
&= \xi_{ij}(\mathbf{h}) - \xi_{i\ell}(\mathbf{h}) - \xi_{kj}(\mathbf{h}) + \xi_{k\ell}(\mathbf{h}),
\end{aligned}$$

and, reordering the summands, the desired equality is obtained.

Finally, to see that the clr cross-covariance matrix is singular for all $\mathbf{h}$, we need to prove that $\boldsymbol{\Xi}(\mathbf{h})\mathbf{j} = \mathbf{0}$ or, equivalently, that the sum of the terms of each row of the clr cross-covariance matrix is zero:

$$\begin{aligned}
\sum_{j=1}^{D} \xi_{ij}(\mathbf{h}) &= \sum_{j=1}^{D} \operatorname{Cov}\left[\ln\left(\frac{Z_i(\mathbf{u})}{g(\mathbf{Z}(\mathbf{u}))}\right), \ln\left(\frac{Z_j(\mathbf{u}+\mathbf{h})}{g(\mathbf{Z}(\mathbf{u}+\mathbf{h}))}\right)\right] \\
&= \operatorname{Cov}\left[\ln\left(\frac{Z_i(\mathbf{u})}{g(\mathbf{Z}(\mathbf{u}))}\right), \sum_{j=1}^{D} \ln\left(\frac{Z_j(\mathbf{u}+\mathbf{h})}{g(\mathbf{Z}(\mathbf{u}+\mathbf{h}))}\right)\right] \\
&= \operatorname{Cov}\left[\ln\left(\frac{Z_i(\mathbf{u})}{g(\mathbf{Z}(\mathbf{u}))}\right), \ln\left(\frac{\prod_{j=1}^{D} Z_j(\mathbf{u}+\mathbf{h})}{(\prod_{j=1}^{D} Z_j(\mathbf{u}+\mathbf{h}))^{\frac{1}{D}D}}\right)\right] \\
&= \operatorname{Cov}\left[\ln\left(\frac{Z_i(\mathbf{u})}{g(\mathbf{Z}(\mathbf{u}))}\right), 0\right],
\end{aligned}$$

which is, as was to be proved, equal to zero. □

As can be deduced from properties 3.6.1 and 3.6.2, the clr cross-covariances reflect asymmetry of the spatial covariance structure, if present. The spatial covariance structure also can be expressed, according to property 3.6.4, in terms of the clr cross-covariances. Therefore the clr cross-covariances are equivalent to alr cross-covariances as long as interest lies exclusively in spatial-structure analysis. Symmetry in the components—the reason given by Aitchison (1986, pp. 78–79) for consideration of the clr transformation for statistical analysis of compositions—proves to be a disadvantage for the application of cokriging to regionalized compositions because of the singularity of the associated matrix. The reason is not so much that standard geostatistical software does not include generalized inverses, a problem that could easily be solved, but rather that modeling algorithms for the cross-covariance functions do not take this constraint into account.

Definition 3.7 *The* clr *cross-covariance of two components* $Z_i(\mathbf{u})$ *and* $Z_j(\mathbf{u})$, $i \neq j$, *of* $\mathbf{Z}(\mathbf{u})$ *is* symmetric *over* $\bar{H}$ *if, for every* $\mathbf{h} \in \bar{H}$,

$$\xi_{ij}(\mathbf{h}) = \xi_{ji}(\mathbf{h});$$

accordingly, the clr *cross-covariance matrix is symmetric over* $\bar{H}$ *if, for every* $\mathbf{h} \in \bar{H}$,

$$\Xi(\mathbf{h}) = \Xi'(\mathbf{h}).$$

As for alr cross-covariances, the concept of symmetry in the indices of the clr cross-covariances corresponds to the usual concept of symmetry of a matrix for the clr cross-covariance matrix for every lag $\mathbf{h} \in \bar{H}$. Furthermore, as can be inferred easily from the definition and property 3.6.3, here again symmetry in the indices is equivalent to symmetry in direction.

Property 3.7 *For every lag* $\mathbf{h}$,

$$\xi_{ij}(\mathbf{h}) = \xi_{ji}(\mathbf{h}) \iff \xi_{ij}(\mathbf{h}) = \xi_{ij}(-\mathbf{h});$$
$$\Xi(\mathbf{h}) = \Xi'(\mathbf{h}) \iff \Xi(\mathbf{h}) = \Xi(-\mathbf{h}).$$

3.1.6 Relationships between specifications

All previous discussions argue for the alr cross-covariances as the best way to completely describe the spatial covariance structure of an r-composition (see definition 3.1): only $(D-1)^2$ functions are needed for this purpose, compared to the initial D^4, and they preserve asymmetry, if present; the number of functions reduces to $(D-1)D/2$ if the spatial covariance structure is symmetric; and the alr cross-covariance matrix is not singular. This is particularly interesting with regard to the application of usual estimation procedures. Nevertheless, the question arises whether the restriction to that specification relinquishes information contained in other specifications. The relations between the different specifications for the spatial covariance structure confirm that this is not the case. A simple conversion places required information immediately at one's disposal. This is significant for spatial-structure analysis as well as for extending the criteria for null spatial correlation of compositions proposed in Chapter 4.

Analogous relations for the nonregionalized case were addressed by Aitchison (1986, pp. 82–83). Again, differences are derived from the possible asymmetry of the spatial covariance structure.

Notation. Hereafter, as is usual, a point replacing a subindex indicates summation over that index and a bar indicates posterior division by the number of elements in the sum, for example

$$\bar{\tau}_{i\cdot}(\mathbf{h}) = \frac{1}{D}\sum_{j=1}^{D}\tau_{i\cdot j}(\mathbf{h}), \qquad \bar{\bar{\tau}}_{\cdot\cdot}(\mathbf{h}) = \frac{1}{D^2}\sum_{i=1}^{D}\sum_{j=1}^{D}\tau_{i\cdot j}(\mathbf{h}),$$

$$\bar{\sigma}_{\cdot j}(\mathbf{h}) = \frac{1}{D}\sum_{i=1}^{D}\sigma_{ij}(\mathbf{h}), \qquad \bar{\bar{\sigma}}_{\cdot\cdot}(\mathbf{h}) = \frac{1}{D^2}\sum_{i=1}^{D}\sum_{j=1}^{D}\sigma_{ij}(\mathbf{h}),$$

and the convention for other equivalent abbreviations is

$$\sigma_{Dj}(\mathbf{h}) = \sigma_{iD}(\mathbf{h}) = \sigma_{DD}(\mathbf{h}) = 0. \tag{3.8}$$

In addition, the following matrices are used:

$$\mathbf{H} = \mathbf{I} + \mathbf{J} = \begin{pmatrix} 2 & 1 & \cdots & 1 \\ 1 & 2 & \cdots & 1 \\ \vdots & \vdots & \ddots & \vdots \\ 1 & 1 & \cdots & 2 \end{pmatrix}, \tag{3.9}$$

with size $(D-1)\times(D-1)$,

$$\operatorname{diag}(\boldsymbol{\Xi}(\mathbf{h})) = \begin{pmatrix} \xi_{11}(\mathbf{h}) & 0 & \cdots & 0 \\ 0 & \xi_{22}(\mathbf{h}) & \cdots & 0 \\ \vdots & \vdots & \ddots & \vdots \\ 0 & 0 & \cdots & \xi_{DD}(\mathbf{h}) \end{pmatrix} \tag{3.10}$$

and

$$\widetilde{\boldsymbol{\Sigma}}(\mathbf{h}) = \begin{pmatrix} \boldsymbol{\Sigma}(\mathbf{h}) & \mathbf{0} \\ \mathbf{0}' & 0 \end{pmatrix} = \begin{pmatrix} \sigma_{11}(\mathbf{h}) & \sigma_{12}(\mathbf{h}) & \cdots & \sigma_{1,D-1}(\mathbf{h}) & 0 \\ \sigma_{21}(\mathbf{h}) & \sigma_{22}(\mathbf{h}) & \cdots & \sigma_{2,D-1}(\mathbf{h}) & 0 \\ \vdots & \vdots & \ddots & \vdots & \vdots \\ \sigma_{D-1,1}(\mathbf{h}) & \sigma_{D-1,2}(\mathbf{h}) & \cdots & \sigma_{D-1,D-1}(\mathbf{h}) & 0 \\ 0 & 0 & \cdots & 0 & 0 \end{pmatrix}. \tag{3.11}$$

Matrices $\mathbf{F} = (\mathbf{I} : -\mathbf{j})$ and $\mathbf{G} = (\mathbf{I} - D^{-1}\mathbf{J})$ from definitions 2.11 and 2.12 are also used, as well as a definition analogous to 3.10 for $\operatorname{diag}(\widetilde{\boldsymbol{\Sigma}}(\mathbf{h}))$. Hereafter, for any matrix $\mathbf{M}$, $\operatorname{diag}(\mathbf{M})$ will denote the diagonal matrix with the same diagonal elements as $\mathbf{M}$.

Property 3.8 *The following relationships hold between the different matrices and covariance functions which can be used to specify the spatial covariance structure:*

1. *between* $\mathbf{T}(\mathbf{h})$ *and* $\boldsymbol{\Sigma}(\mathbf{h}) + \boldsymbol{\Sigma}'(\mathbf{h})$:

$$\begin{aligned}\boldsymbol{\Sigma}(\mathbf{h}) + \boldsymbol{\Sigma}'(\mathbf{h}) &= -\mathbf{F}\mathbf{T}(\mathbf{h})\mathbf{F}', \\ \sigma_{ij}(\mathbf{h}) + \sigma_{ji}(\mathbf{h}) &= \tau_{i\cdot D}(\mathbf{h}) + \tau_{j\cdot D}(\mathbf{h}) - \tau_{i\cdot j}(\mathbf{h});\end{aligned}$$

2. *between* $\mathbf{T}(\mathbf{h})$ *and* $\boldsymbol{\Xi}(\mathbf{h}) + \boldsymbol{\Xi}'(\mathbf{h})$:

$$\begin{aligned}\boldsymbol{\Xi}(\mathbf{h}) + \boldsymbol{\Xi}'(\mathbf{h}) &= -\mathbf{G}\mathbf{T}(\mathbf{h})\mathbf{G}', \\ \xi_{ij}(\mathbf{h}) + \xi_{ji}(\mathbf{h}) &= \bar{\tau}_{i\cdot}(\mathbf{h}) + \bar{\tau}_{j\cdot}(\mathbf{h}) - \tau_{i\cdot j}(\mathbf{h}) - \bar{\bar{\tau}}_{\cdot\cdot}(\mathbf{h});\end{aligned}$$

3. *between* $\boldsymbol{\Sigma}(\mathbf{h})$ *and* $\mathbf{T}(\mathbf{h})$:

$$\begin{aligned}\mathbf{T}(\mathbf{h}) &= \mathbf{J}\mathrm{diag}(\widetilde{\boldsymbol{\Sigma}}(\mathbf{h})) + \mathrm{diag}(\widetilde{\boldsymbol{\Sigma}}(\mathbf{h}))\mathbf{J} - \widetilde{\boldsymbol{\Sigma}}(\mathbf{h}) - \widetilde{\boldsymbol{\Sigma}}'(\mathbf{h}), \\ \tau_{i\cdot j}(\mathbf{h}) &= \sigma_{ii}(\mathbf{h}) + \sigma_{jj}(\mathbf{h}) - \sigma_{ij}(\mathbf{h}) - \sigma_{ji}(\mathbf{h});\end{aligned}$$

4. *between* $\boldsymbol{\Sigma}(\mathbf{h})$ *and* $\boldsymbol{\Xi}(\mathbf{h})$:

$$\begin{aligned}\boldsymbol{\Xi}(\mathbf{h}) &= -\mathbf{F}'\mathbf{H}^{-1}\boldsymbol{\Sigma}(\mathbf{h})\mathbf{H}^{-1}\mathbf{F}, \\ \xi_{ij}(\mathbf{h}) &= \sigma_{ij}(\mathbf{h}) - \bar{\sigma}_{i\cdot}(\mathbf{h}) - \bar{\sigma}_{\cdot j}(\mathbf{h}) + \bar{\bar{\sigma}}_{\cdot\cdot}(\mathbf{h});\end{aligned}$$

5. *between* $\boldsymbol{\Xi}(\mathbf{h})$ *and* $\mathbf{T}(\mathbf{h})$:

$$\begin{aligned}\mathbf{T}(\mathbf{h}) &= \mathbf{J}\mathrm{diag}(\boldsymbol{\Xi}(\mathbf{h})) + \mathrm{diag}(\boldsymbol{\Xi}(\mathbf{h}))\mathbf{J} - \boldsymbol{\Xi}(\mathbf{h}) - \boldsymbol{\Xi}'(\mathbf{h}), \\ \tau_{i\cdot j}(\mathbf{h}) &= \xi_{ii}(\mathbf{h}) + \xi_{jj}(\mathbf{h}) - \xi_{ij}(\mathbf{h}) - \xi_{ji}(\mathbf{h});\end{aligned}$$

6. *between* $\boldsymbol{\Xi}(\mathbf{h})$ *and* $\boldsymbol{\Sigma}(\mathbf{h})$:

$$\begin{aligned}\boldsymbol{\Sigma}(\mathbf{h}) &= \mathbf{F}\boldsymbol{\Xi}(\mathbf{h})\mathbf{F}', \\ \sigma_{ij}(\mathbf{h}) &= \xi_{ij}(\mathbf{h}) - \xi_{iD}(\mathbf{h}) - \xi_{Dj}(\mathbf{h}) + \xi_{DD}(\mathbf{h}).\end{aligned}$$

Proof. We shall prove here only property 3.8.1 as an example of how to proceed; the other properties are proven in an analogous manner. Consider first the matrix formulation of the property $\boldsymbol{\Sigma}(\mathbf{h}) + \boldsymbol{\Sigma}'(\mathbf{h}) = -\mathbf{F}\mathbf{T}(\mathbf{h})\mathbf{F}'$. Note that the left-hand side corresponds to $\sigma_{ij}(\mathbf{h}) + \sigma_{ji}(\mathbf{h})$ for each element of the resulting sum, while on the right-hand side we have the product of three matrices with negative sign. Multiplication of

$$\mathbf{F} = -\begin{pmatrix} 1 & 0 & \cdots & 0 & -1 \\ 0 & 1 & \cdots & 0 & -1 \\ \vdots & \vdots & \ddots & \vdots & \vdots \\ 0 & 1 & \cdots & 1 & -1 \end{pmatrix}$$

by

$$\mathbf{T}(\mathbf{h}) = \begin{pmatrix} \tau_{1\cdot 1}(\mathbf{h}) & \tau_{1\cdot 2}(\mathbf{h}) & \cdots & \tau_{1\cdot D}(\mathbf{h}) \\ \tau_{2\cdot 1}(\mathbf{h}) & \tau_{2\cdot 2}(\mathbf{h}) & \cdots & \tau_{2\cdot D}(\mathbf{h}) \\ \vdots & \vdots & \ddots & \vdots \\ \tau_{(D-1)\cdot 1}(\mathbf{h}) & \tau_{(D-1)\cdot 2}(\mathbf{h}) & \cdots & \tau_{(D-1)\cdot D}(\mathbf{h}) \\ \tau_{D\cdot 1}(\mathbf{h}) & \tau_{D\cdot 2}(\mathbf{h}) & \cdots & \tau_{D\cdot D}(\mathbf{h}) \end{pmatrix}$$

yields

$$\begin{pmatrix} \tau_{1\cdot 1}(\mathbf{h}) - \tau_{D\cdot 1}(\mathbf{h}) & \tau_{1\cdot 2}(\mathbf{h}) - \tau_{D\cdot 2}(\mathbf{h}) & \cdots & \tau_{1\cdot D}(\mathbf{h}) - \tau_{D\cdot D}(\mathbf{h}) \\ \tau_{2\cdot 1}(\mathbf{h}) - \tau_{D\cdot 1}(\mathbf{h}) & \tau_{2\cdot 2}(\mathbf{h}) - \tau_{D\cdot 2}(\mathbf{h}) & \cdots & \tau_{2\cdot D}(\mathbf{h}) - \tau_{D\cdot D}(\mathbf{h}) \\ \vdots & \vdots & \ddots & \vdots \\ \tau_{(D-1)\cdot 1}(\mathbf{h}) - \tau_{D\cdot 1}(\mathbf{h}) & \tau_{(D-1)\cdot 2}(\mathbf{h}) - \tau_{D\cdot 2}(\mathbf{h}) & \cdots & \tau_{(D-1)\cdot D}(\mathbf{h}) - \tau_{D\cdot D}(\mathbf{h}) \end{pmatrix}$$

and after postmultiplication by

$$\mathbf{F}' = \begin{pmatrix} 1 & 0 & \cdots & 0 \\ 0 & 1 & \cdots & 0 \\ \vdots & \vdots & \ddots & \vdots \\ -1 & -1 & \cdots & -1 \end{pmatrix}$$

and changing the sign, the result for each element is

$$-\tau_{i\cdot j}(\mathbf{h}) + \tau_{D\cdot j}(\mathbf{h}) + \tau_{i\cdot D}(\mathbf{h}) - \tau_{D\cdot D}(\mathbf{h}).$$

As by property 3.3.2, $\tau_{D\cdot j}(\mathbf{h}) = \tau_{j\cdot D}(\mathbf{h})$ and by property 3.3.1, $\tau_{D\cdot D}(\mathbf{h}) = 0$, the expected element expression given above is obtained. Thus, we need to prove that $\sigma_{ij}(\mathbf{h}) + \sigma_{ji}(\mathbf{h}) = \tau_{i\cdot D}(\mathbf{h}) + \tau_{j\cdot D}(\mathbf{h}) - \tau_{i\cdot j}(\mathbf{h})$. To do that, we note that using definition 3.4 first, then properties 3.1.6, 3.1.4, 3.1.5, 3.1.6, 3.1.5 in the given order, and finally using definition 3.3, the following sequence of equalities holds.

$$\begin{aligned} &\sigma_{ij}(\mathbf{h}) + \sigma_{ji}(\mathbf{h}) \\ &\quad= \sigma_{ij\cdot DD}(\mathbf{h}) + \sigma_{ji\cdot DD}(\mathbf{h}) \\ &\quad= \sigma_{ij\cdot Di}(\mathbf{h}) + \sigma_{ii\cdot DD}(\mathbf{h}) + \sigma_{Dj\cdot Di}(\mathbf{h}) + \sigma_{Di\cdot DD}(\mathbf{h}) \\ &\qquad+ \sigma_{ji\cdot Dj}(\mathbf{h}) + \sigma_{jj\cdot DD}(\mathbf{h}) + \sigma_{Di\cdot Dj}(\mathbf{h}) + \sigma_{Dj\cdot DD}(\mathbf{h}) \end{aligned}$$

$$\begin{aligned}
&= \sigma_{ii\cdot DD}(\mathbf{h}) + \sigma_{jj\cdot DD}(\mathbf{h}) - \sigma_{ii\cdot Dj}(\mathbf{h}) + \sigma_{ji\cdot Dj}(\mathbf{h}) \\
&= \sigma_{ii\cdot DD}(\mathbf{h}) + \sigma_{jj\cdot DD}(\mathbf{h}) - \sigma_{ii\cdot jj}(\mathbf{h}) - \sigma_{ij\cdot jj}(\mathbf{h}) \\
&\quad - \sigma_{ji\cdot Dj}(\mathbf{h}) - \sigma_{jj\cdot Dj}(\mathbf{h}) + \sigma_{ji\cdot Dj}(\mathbf{h}) \\
&= \sigma_{ii\cdot DD}(\mathbf{h}) + \sigma_{jj\cdot DD}(\mathbf{h}) - \sigma_{ii\cdot jj}(\mathbf{h}) - \sigma_{ji\cdot Dj}(\mathbf{h}) + \sigma_{ji\cdot Dj}(\mathbf{h}) \\
&= \tau_{i\cdot D}(\mathbf{h}) + \tau_{j\cdot D}(\mathbf{h}) - \tau_{i\cdot j}(\mathbf{h}).
\end{aligned}$$

Thus the desired equality is proven.

The other properties are proven in an analogous manner, as mentioned. Starting from the definition, reduce the involved covariances to covariances in terms of $\sigma_{ij\cdot k\ell}(\mathbf{h})$, use property 3.1 together with basic properties of the covariance and logarithmic functions and, finally, use the notation introduced at the beginning of this subsection to arrive at the required expressions. □

Relationships 1 and 2 of property 3.8 show that $\mathbf{T}(\mathbf{h})$, as expected, is sufficient for complete determination of the other specifications only if the corresponding conditions of symmetry are satisfied; but it is always possible, as evident from relationships 3 and 5, to derive each lr autocovariance from one of the other two specifications.

The interest in these expressions is not only theoretical: their matrix formulation is important for computational algorithms, if an easy change from one representation to the other is required.

3.1.7 Symmetry of the spatial covariance structure

As expected, the strong dependence between the different specifications for the spatial covariance structure implies equivalence between the different definitions of symmetry. Symmetry of the cross-covariance matrix of the log-transformed r-composition, $\ln(\mathbf{Z}(\mathbf{u}))$, is a sufficient condition for this equivalency.

Property 3.9 *For every lag* $\mathbf{h}$ *the following properties are equivalent:*

1. $\sigma_{ij\cdot k\ell}(\mathbf{h}) = \sigma_{ji\cdot \ell k}(\mathbf{h})$, *for all* $i, j, k, \ell \in \{1, 2, \ldots, D\}$,

2. $\sigma_{ij}(\mathbf{h}) = \sigma_{ji}(\mathbf{h})$, *for all* $i, j \in \{1, 2, \ldots, D-1\}$,

3. $\xi_{ij}(\mathbf{h}) = \xi_{ji}(\mathbf{h})$ *for all* $i, j \in \{1, 2, \ldots, D\}$.

Proof. If, for all $i, j, k, \ell \in \{1, 2, \ldots, D\}$, $\sigma_{ij\cdot k\ell}(\mathbf{h}) = \sigma_{ji\cdot \ell k}(\mathbf{h})$, it follows from the definition of $\sigma_{ij}(\mathbf{h})$ that

$$\sigma_{ij}(\mathbf{h}) = \sigma_{ij\cdot DD}(\mathbf{h}) = \sigma_{ji\cdot DD}(\mathbf{h}) = \sigma_{ji}(\mathbf{h}),$$

and thus 1 implies 2. Conversely, if $\sigma_{ij}(\mathbf{h}) = \sigma_{ji}(\mathbf{h})$ for all $i, j \in \{1, 2, \dots, D\}$, then, using property 3.4.4, the symmetry of the spatial covariance structure is obtained. Furthermore, the relation

$$\sigma_{ij}(\mathbf{h}) = \sigma_{ji}(\mathbf{h}) \Longleftrightarrow \xi_{ij}(\mathbf{h}) = \xi_{ji}(\mathbf{h})$$

results from properties 3.8.4 and 3.8.6. □

The equivalence of the symmetry of the different cross-covariance functions stated in property 3.9 justifies the already-used term *symmetric spatial covariance structure* to refer to the symmetry of any of the cross-covariances which appear in it.

Property 3.10 *For every lag* $\mathbf{h}$

$$\left\{ \begin{array}{c} \eta_{ij}(\mathbf{h}) = \eta_{ji}(\mathbf{h}) \\ \forall i, j \in \{1, 2, \dots, D\} \end{array} \right\} \Longrightarrow \left\{ \begin{array}{c} \sigma_{ij \cdot k\ell}(\mathbf{h}) = \sigma_{ji \cdot \ell k}(\mathbf{h}) \\ \forall i, j, k, \ell \in \{1, 2, \dots, D\} \end{array} \right\}$$

where $\eta_{ij}(\mathbf{h}) = \mathrm{Cov}\left[\ln(Z_i(\mathbf{u})), \ln(Z_j(\mathbf{u}+\mathbf{h}))\right]$.

Proof. Definition 3.1 yields

$$\sigma_{ij \cdot k\ell}(\mathbf{h}) = \eta_{ij}(\mathbf{h}) + \eta_{k\ell}(\mathbf{h}) - \eta_{i\ell}(\mathbf{h}) - \eta_{kj}(\mathbf{h})$$

and, given that $\eta_{ij}(\mathbf{h}) = \eta_{ji}(\mathbf{h})$ for all $i, j \in \{1, 2, \dots, D\}$, the symmetry of the spatial covariance structure follows. □

3.2 Spatial covariance structure under intrinsic hypothesis

In this section, a D-part lr intrinsic r-composition $\mathbf{Z}(\mathbf{u})$, $\mathbf{u} \in \Omega$, is considered, except in Subsection 3.2.5. There, to establish relationships between the general and intrinsic spatial covariance structure, second-order lr stationarity of the r-composition is required.

3.2.1 Intrinsic spatial covariance structure

Definition 3.8 *The* intrinsic spatial covariance structure *of an r-composition* $\mathbf{Z}(\mathbf{u})$ *is, for* $i, j, k, \ell \in \{1, 2, \dots, D\}$*, the set of functions*

$$\gamma_{ij \cdot k\ell}(\mathbf{h}) = \tfrac{1}{2}\mathrm{Cov}\left[\ln\left(\frac{Z_i(\mathbf{u})}{Z_k(\mathbf{u})}\right) - \ln\left(\frac{Z_i(\mathbf{u}+\mathbf{h})}{Z_k(\mathbf{u}+\mathbf{h})}\right), \ln\left(\frac{Z_j(\mathbf{u})}{Z_l(\mathbf{u})}\right) - \ln\left(\frac{Z_j(\mathbf{u}+\mathbf{h})}{Z_l(\mathbf{u}+\mathbf{h})}\right)\right].$$

Property 3.11 *The intrinsic spatial covariance structure of* $\mathbf{Z}(\mathbf{u})$ *has the following properties for* $i, j, k, \ell \in \{1, 2, \ldots, D\}$:

1. $\gamma_{ij\cdot k\ell}(0) = 0$,

2. $\gamma_{ij\cdot k\ell}(\mathbf{h}) = \gamma_{ji\cdot \ell k}(\mathbf{h}) = \gamma_{ij\cdot k\ell}(-\mathbf{h})$,

3. $\gamma_{ij\cdot i\ell}(\mathbf{h}) = \gamma_{ij\cdot kj}(\mathbf{h}) = \gamma_{ij\cdot ij}(\mathbf{h}) = \gamma_{ii\cdot ii}(\mathbf{h}) = 0$,

4. $\gamma_{ij\cdot k\ell}(\mathbf{h}) = -\gamma_{kj\cdot i\ell}(\mathbf{h}) = \gamma_{k\ell\cdot ij}(\mathbf{h}) = -\gamma_{i\ell\cdot kj}(\mathbf{h})$,

5. $|\gamma_{ij\cdot k\ell}(\mathbf{h})| \leq \sqrt{\gamma_{ii\cdot kk}(\mathbf{h})}\,\sqrt{\gamma_{jj\cdot \ell\ell}(\mathbf{h})}$ *(Cauchy-Schwarz inequality).*

The intrinsic spatial covariance structure is symmetrical both in $\mathbf{h}$ and in the indices. This means that it is impossible to account for any possible directional differences in the spatial dependence among the components of the r-composition.

As for the spatial covariance structure in the second-order stationary case, because of the characteristics described in property 3.11, there is a large decrease in the number of the D^4 possible functions $\gamma_{ij\cdot k\ell}(\mathbf{h})$ that are not interdependent. To reduce this number further, different specifications for the intrinsic spatial covariance structure will be analyzed in a completely analogous way. At the same time, the adequacy of these specifications for cokriging and for the necessary process of modeling of the experimental semivariograms and cross-semivariograms will be examined.

3.2.2 lr semivariogram

Definition 3.9 *The* lr semivariogram *of any two components* $Z_i(\mathbf{u})$, $Z_j(\mathbf{u})$, $i, j = 1, 2, \ldots, D$, *of an r-composition* $\mathbf{Z}(\mathbf{u})$ *is the semivariogram of the corresponding logratio,*

$$\gamma_{i\cdot j}(\mathbf{h}) = \gamma_{ii\cdot jj}(\mathbf{h}),$$

and the intrinsic variation matrix *is the matrix of* lr *semivariograms with dimension* $D \times D$,

$$\mathbf{\Gamma}(\mathbf{h}) = (\,\gamma_{i\cdot j}(\mathbf{h})\,)\,.$$

Property 3.12 *The* lr *semivariograms have the following properties for* i, j, $k, \ell \in \{1, 2, \ldots, D\}$:

1. $\gamma_{i\cdot i}(\mathbf{h}) = 0$,

2. $\mathbf{\Gamma}(0) = \mathbf{0}$,

3. $\mathbf{\Gamma}(\mathbf{h}) = \mathbf{\Gamma}'(\mathbf{h}) = \mathbf{\Gamma}(-\mathbf{h})$,

4. $2\gamma_{ij\cdot k\ell}(\mathbf{h}) = \gamma_{i\cdot\ell}(\mathbf{h}) + \gamma_{j\cdot k}(\mathbf{h}) - \gamma_{i\cdot j}(\mathbf{h}) - \gamma_{k\cdot\ell}(\mathbf{h})$,

5. Cauchy-Schwarz inequality:

(a) $|\gamma_{i\cdot\ell}(\mathbf{h}) + \gamma_{j\cdot k}(\mathbf{h}) - \gamma_{i\cdot j}(\mathbf{h}) - \gamma_{k\cdot\ell}(\mathbf{h})| \leq 2\sqrt{\gamma_{i\cdot k}(\mathbf{h})}\sqrt{\gamma_{j\cdot\ell}(\mathbf{h})}$,

(b) $|\gamma_{i\cdot k}(\mathbf{h}) + \gamma_{j\cdot k}(\mathbf{h}) - \gamma_{i\cdot j}(\mathbf{h})| \leq 2\sqrt{\gamma_{i\cdot k}(\mathbf{h})}\sqrt{\gamma_{j\cdot k}(\mathbf{h})}$,

(c) $|\gamma_{i\cdot D}(\mathbf{h}) + \gamma_{j\cdot D}(\mathbf{h}) - \gamma_{i\cdot j}(\mathbf{h})| \leq 2\sqrt{\gamma_{i\cdot D}(\mathbf{h})}\sqrt{\gamma_{j\cdot D}(\mathbf{h})}$.

Proof. The proof is direct using the corresponding definition and the properties of the intrinsic spatial covariance structure stated in property 3.11, taking into account that property 3.12.5a is derived from property 3.12.4 and the definition, while properties 3.12.5b and 3.12.5c are just particular cases of property 3.12.5a. □

The $(D-1)D/2$ functionally independent lr semivariograms describe the intrinsic spatial covariance structure completely; furthermore, the experimental lr semivariograms—or semivariograms of all possible pairwise logratios among the components of $\mathbf{Z}(\mathbf{u})$—can be calculated and modeled by means of existing programs. This supports the choice of lr semivariograms as the best tool for intrinsic spatial-structure analysis. An extended discussion of how to find valid models is given in Section 6.2.

3.2.3 alr cross-semivariogram

Definition 3.10 *For two components* $Z_i(\mathbf{u})$, $Z_j(\mathbf{u})$, $i, j = 1, 2, \ldots, D$, *of an r-composition* $\mathbf{Z}(\mathbf{u})$, *the* alr cross-semivariogram *is the function*

$$\psi_{ij}(\mathbf{h}) = \gamma_{ij\cdot DD}(\mathbf{h});$$

the alr semivariogram *of any component is included as a particular case for* $i = j$. *The* intrinsic alr cross-covariance matrix *of* $\mathbf{Z}(\mathbf{u})$ *is then the matrix, with dimension* $(D-1) \times (D-1)$, *of the* alr *cross-semivariograms for every possible pair of components,*

$$\boldsymbol{\Psi}(\mathbf{h}) = (\, \psi_{ij}(\mathbf{h})\,).$$

Property 3.13 alr *cross-semivariograms have the following properties* $(i, j, k, \ell \in \{1, 2, \ldots, D-1\})$:

1. $\boldsymbol{\Psi}(0) = \mathbf{0}$,

2. $\boldsymbol{\Psi}(\mathbf{h}) = \boldsymbol{\Psi}'(\mathbf{h}) = \boldsymbol{\Psi}(-\mathbf{h})$,

3. $\gamma_{ij\cdot k\ell}(\mathbf{h}) = \psi_{ij}(\mathbf{h}) + \psi_{k\ell}(\mathbf{h}) - \psi_{i\ell}(\mathbf{h}) - \psi_{kj}(\mathbf{h})$,

4. $|\psi_{ij}(\mathbf{h}) + \psi_{k\ell}(\mathbf{h}) - \psi_{i\ell}(\mathbf{h}) - \psi_{kj}(\mathbf{h})|$
$\leq \sqrt{\psi_{ii}(\mathbf{h}) + \psi_{kk}(\mathbf{h}) - 2\psi_{ik}(\mathbf{h})}\sqrt{\psi_{jj}(\mathbf{h}) + \psi_{\ell\ell}(\mathbf{h}) - 2\psi_{j\ell}(\mathbf{h})}$,

5. $|\psi_{ij}(\mathbf{h})| \leq \sqrt{\psi_{ii}(\mathbf{h})}\,\sqrt{\psi_{jj}(\mathbf{h})}$.

As can be seen from property 3.13.3, the alr cross-semivariograms, like the lr semivariograms, completely specify the intrinsic spatial covariance structure. Also, the total number of functions that are not interdependent is $(D-1)D/2$. Only the asymmetry of the components argues against this specification, while the difficulty in adjustment of a model to experimental cross-semivariograms mentioned by Myers (1982) can be overcome with the approach given by Yao and Journel (1998) and addressed briefly in Section 6.2.

3.2.4 clr cross-semivariogram

The clr cross-semivariogram is introduced here to complete the presentation because the intrinsic clr cross-covariance matrix is also singular, as is the clr cross-covariance matrix.

Definition 3.11 *For two components $Z_i(\mathbf{u})$, $Z_j(\mathbf{u})$, $i, j = 1, 2, \ldots, D$, of an r-composition $\mathbf{Z}(\mathbf{u})$, the* clr cross-semivariogram *is the function*

$$\delta_{ij}(\mathbf{h}) = \operatorname{Cov}\left[(u_i(\mathbf{u}) - u_i(\mathbf{u}+\mathbf{h})), (u_j(\mathbf{u}) - u_j(\mathbf{u}+\mathbf{h}))\right],$$

where

$$u_i(\mathbf{u}) = \ln\left(\frac{Z_i(\mathbf{u})}{g(\mathbf{Z}(\mathbf{u}))}\right).$$

The clr semivariogram *of any component is included as a particular case for $i = j$. The* intrinsic clr cross-covariance matrix, *with dimension $D \times D$, is then the matrix of the* clr *cross-semivariograms for any possible pair of components,*

$$\boldsymbol{\Delta}(\mathbf{h}) = (\,\delta_{ij}(\mathbf{h})\,).$$

Property 3.14 clr *cross-semivariograms have the following properties ($i, j, k, \ell \in \{1, 2, \ldots, D\}$):*

1. $\boldsymbol{\Delta}(0) = \mathbf{0}$,

2. $\boldsymbol{\Delta}(\mathbf{h}) = \boldsymbol{\Delta}'(\mathbf{h}) = \boldsymbol{\Delta}(-\mathbf{h})$,

3. $\gamma_{ij\cdot k\ell}(\mathbf{h}) = \delta_{ij}(\mathbf{h}) + \delta_{k\ell}(\mathbf{h}) - \delta_{i\ell}(\mathbf{h}) - \delta_{kj}(\mathbf{h})$,

4. $\boldsymbol{\Delta}(\mathbf{h})\mathbf{j} = \mathbf{0}$,

5. Cauchy-Schwarz inequality:

$$|\delta_{ij}(\mathbf{h}) + \delta_{k\ell}(\mathbf{h}) - \delta_{i\ell}(\mathbf{h}) - \delta_{kj}(\mathbf{h})|$$
$$\leq \sqrt{\delta_{ii}(\mathbf{h}) + \delta_{kk}(\mathbf{h}) - 2\delta_{ik}(\mathbf{h})}\sqrt{\delta_{jj}(\mathbf{h}) + \delta_{\ell\ell}(\mathbf{h}) - 2\delta_{j\ell}(\mathbf{h})}\,.$$

Property 3.14.4 implies that the intrinsic clr cross-covariance matrix is singular.

3.2.5 Further relationships between specifications

Relationships between the different possible specifications for the spatial covariance structure of an r-composition were discussed in Section 3.1.6. Here, corresponding equations for the different specifications for the intrinsic spatial covariance structure also arise. In this instance, however, no conditions of asymmetry have to be observed. Therefore, the equations for every $\mathbf{h}$ are analogous to those for the nonregionalized case (see Aitchison 1986, pp. 82–83) and are not all listed here. An exception is made for the equalities that follow because lr semivariograms and alr cross-semivariograms are both of special interest for further analysis, chiefly for estimation and modeling purposes.

As in equations 3.8 and 3.11 in Section 3.1.6, the following conventions are used:

$$\psi_{iD}(\mathbf{h}) = \psi_{Di}(\mathbf{h}) = \psi_{DD}(\mathbf{h}) = 0,$$

and

$$\widetilde{\boldsymbol{\Psi}}(\mathbf{h}) = \begin{pmatrix} \boldsymbol{\Psi}(\mathbf{h}) & \mathbf{0} \\ \mathbf{0}' & 0 \end{pmatrix} = \begin{pmatrix} \psi_{11}(\mathbf{h}) & \psi_{12}(\mathbf{h}) & \cdots & \psi_{1,D-1}(\mathbf{h}) & 0 \\ \psi_{21}(\mathbf{h}) & \psi_{22}(\mathbf{h}) & \cdots & \psi_{2,D-1}(\mathbf{h}) & 0 \\ \vdots & \vdots & \ddots & \vdots & \vdots \\ \psi_{D-1,1}(\mathbf{h}) & \psi_{D-1,2}(\mathbf{h}) & \cdots & \psi_{D-1,D-1}(\mathbf{h}) & 0 \\ 0 & 0 & \cdots & 0 & 0 \end{pmatrix},$$

as well as matrix $\mathbf{F} = (\mathbf{I} : -\mathbf{j})$ from definition 2.11.

Property 3.15 *For every lag* $\mathbf{h}$*,*

$$\begin{aligned} \psi_{ij}(\mathbf{h}) &= \tfrac{1}{2}\left(\gamma_{i\cdot D}(\mathbf{h}) + \gamma_{j\cdot D}(\mathbf{h}) - \gamma_{i\cdot j}(\mathbf{h})\right), \\ \gamma_{i\cdot j}(\mathbf{h}) &= \psi_{ii}(\mathbf{h}) + \psi_{jj}(\mathbf{h}) - 2\psi_{ij}(\mathbf{h})), \end{aligned}$$

or in matrix notation,

$$\begin{aligned} \boldsymbol{\Psi}(\mathbf{h}) &= -\tfrac{1}{2}\mathbf{F}\boldsymbol{\Gamma}(\mathbf{h})\mathbf{F}', \\ \boldsymbol{\Gamma}(\mathbf{h}) &= \mathbf{J}\mathrm{diag}(\widetilde{\boldsymbol{\Psi}}(\mathbf{h})) + \mathrm{diag}(\widetilde{\boldsymbol{\Psi}}(\mathbf{h}))\mathbf{J} - 2\widetilde{\boldsymbol{\Psi}}(\mathbf{h}). \end{aligned}$$

Property 3.15 reflects the fact that each alr cross-semivariogram is dependent on the semivariograms of the pairwise logratios of the involved components and the component taken as divisor. Furthermore, it should be noted that $\gamma_{i \cdot j}(\mathbf{h})$ is, in fact, the semivariogram of the pairwise differences, which is evident from the fact that the pairwise ratios agree with the corresponding pairwise differences of the alr transformed variables:

$$\ln\left(\frac{Z_i(\mathbf{u})}{Z_j(\mathbf{u})}\right) = \ln\left(\frac{Z_i(\mathbf{u})}{Z_D(\mathbf{u})}\right) - \ln\left(\frac{Z_j(\mathbf{u})}{Z_D(\mathbf{u})}\right).$$

It follows that estimation and modeling of the semivariogram of

$$\ln\left(\frac{Z_i(\mathbf{u})}{Z_D(\mathbf{u})}\right) - \ln\left(\frac{Z_j(\mathbf{u})}{Z_D(\mathbf{u})}\right)$$

is the same as the estimation and modeling of $\gamma_{i \cdot j}(\mathbf{h})$. This is important for modeling purposes and will be addressed in Section 6.2.

In addition, the conditions of dependence between the different specifications of the spatial covariance structure and those corresponding to the intrinsic case are of interest. Assume that all the necessary variances and covariances exist and are finite, i.e., that the r-composition is second-order lr stationary.

Property 3.16 *In general, the following statements hold:*

1. $\gamma_{ij \cdot k\ell}(\mathbf{h}) = \sigma_{ij \cdot k\ell}(0) - \frac{1}{2}(\sigma_{ij \cdot k\ell}(\mathbf{h}) + \sigma_{ji \cdot \ell k}(\mathbf{h}))$,
2. $\mathbf{\Gamma}(\mathbf{h}) = \mathbf{T}(0) - \mathbf{T}(\mathbf{h})$,
3. $\mathbf{\Psi}(\mathbf{h}) = \mathbf{\Sigma}(0) - \frac{1}{2}(\mathbf{\Sigma}(\mathbf{h}) + \mathbf{\Sigma}'(\mathbf{h}))$,
4. $\mathbf{\Delta}(\mathbf{h}) = \mathbf{\Xi}(0) - \frac{1}{2}(\mathbf{\Xi}(\mathbf{h}) + \mathbf{\Xi}'(\mathbf{h}))$.

Property 3.17 *If the spatial covariance structure is symmetric, then*

1. $\gamma_{ij \cdot k\ell}(\mathbf{h}) = \sigma_{ij \cdot k\ell}(0) - \sigma_{ij \cdot k\ell}(\mathbf{h})$,
2. $\mathbf{\Gamma}(\mathbf{h}) = \mathbf{T}(0) - \mathbf{T}(\mathbf{h})$,
3. $\mathbf{\Psi}(\mathbf{h}) = \mathbf{\Sigma}(0) - \mathbf{\Sigma}(\mathbf{h})$,
4. $\mathbf{\Delta}(\mathbf{h}) = \mathbf{\Xi}(0) - \mathbf{\Xi}(\mathbf{h})$.

These relations correspond to those of Matheron (1965) for the relationship between the autocovariance and the semivariogram in the case of second-order stationarity and are derived analogously.

Because $\mathbf{\Gamma}(\mathbf{h})$ and $\mathbf{\Sigma}(\mathbf{h})$ appear particularly suitable for use in estimation procedures and spatial covariance structural analysis, the relationship between both is of special interest. It is summarized as follows:

Property 3.18

$$-\tfrac{1}{2}\,\mathbf{F}\boldsymbol{\Gamma}(\mathbf{h})\mathbf{F}' = \boldsymbol{\Sigma}(0) - \tfrac{1}{2}\,(\boldsymbol{\Sigma}(\mathbf{h}) + \boldsymbol{\Sigma}(\mathbf{h})').$$

Property 3.19 *If the spatial covariance structure is symmetric, then*

$$\boldsymbol{\Sigma}(\mathbf{h}) = \boldsymbol{\Sigma}(0) + \tfrac{1}{2}\,\mathbf{F}\boldsymbol{\Gamma}(\mathbf{h})\mathbf{F}'.$$

The proof of both statements is obtained using properties 3.15, 3.16, and 3.17. Corresponding expressions for $\boldsymbol{\Gamma}(\mathbf{h})$ in terms of $\boldsymbol{\Sigma}(\mathbf{h})$ are deduced using the extended matrix $\widetilde{\boldsymbol{\Sigma}}(\mathbf{h})$, as defined in equation 3.11 and properties 3.8.5 and 3.16.2:

Property 3.20

$$\begin{aligned}\boldsymbol{\Gamma}(\mathbf{h}) &= \mathbf{J}\mathrm{diag}(\widetilde{\boldsymbol{\Sigma}}(0) - \widetilde{\boldsymbol{\Sigma}}(\mathbf{h})) + \mathrm{diag}(\widetilde{\boldsymbol{\Sigma}}(0) - \widetilde{\boldsymbol{\Sigma}}(\mathbf{h}))\mathbf{J}\\ &\quad -(\widetilde{\boldsymbol{\Sigma}}(0) - \widetilde{\boldsymbol{\Sigma}}(\mathbf{h})) - (\widetilde{\boldsymbol{\Sigma}}(0) - \widetilde{\boldsymbol{\Sigma}}(\mathbf{h}))'.\end{aligned}$$

Property 3.21 *If the spatial covariance structure is symmetric, then*

$$\boldsymbol{\Gamma}(\mathbf{h}) = \mathbf{J}\mathrm{diag}(\widetilde{\boldsymbol{\Sigma}}(0) - \widetilde{\boldsymbol{\Sigma}}(\mathbf{h})) + \mathrm{diag}(\widetilde{\boldsymbol{\Sigma}}(0) - \widetilde{\boldsymbol{\Sigma}}(\mathbf{h}))\mathbf{J} - 2(\widetilde{\boldsymbol{\Sigma}}(0) - \widetilde{\boldsymbol{\Sigma}}(\mathbf{h})).$$

3.3 Spatial covariance structure of an r-basis

Let $\mathbf{W}(\mathbf{u})$ be a second-order log-stationary r-basis as defined in Chapter 2 (see definition 2.14).

Definition 3.12 *The* log cross-covariance *of any two components* $w_i(\mathbf{u})$, $w_j(\mathbf{u})$, $i, j = 1, 2, \ldots, D$, *of* $\mathbf{W}(\mathbf{u})$ *is the function:*

$$\omega_{ij}(\mathbf{h}) = \mathrm{Cov}\left[\ln(w_i(\mathbf{u})), \ln(w_j(\mathbf{u}+\mathbf{h}))\right],$$

with associated matrix

$$\boldsymbol{\Omega}(\mathbf{h}) = (\,\omega_{ij}(\mathbf{h})\,)\,.$$

Property 3.22 *The log cross-covariances have the following properties* (i, j, k, $\ell \in \{1, 2, \ldots, D\}$) :

1. *in general,* $\boldsymbol{\Omega}(\mathbf{h}) \neq \boldsymbol{\Omega}'(\mathbf{h})$,
2. *in general,* $\boldsymbol{\Omega}(\mathbf{h}) \neq \boldsymbol{\Omega}(-\mathbf{h})$,
3. $\boldsymbol{\Omega}(\mathbf{h}) = \boldsymbol{\Omega}'(-\mathbf{h})$,

4. $\tau_{i\cdot j}(\mathbf{h}) = \omega_{ii}(\mathbf{h}) + \omega_{jj}(\mathbf{h}) - \omega_{ij}(\mathbf{h}) - \omega_{ji}(\mathbf{h})$, *or in matrix notation,*

   $\mathbf{T}(\mathbf{h}) = \mathbf{J}\mathrm{diag}(\boldsymbol{\Omega}(\mathbf{h})) + \mathrm{diag}(\boldsymbol{\Omega}(\mathbf{h}))\mathbf{J} - \boldsymbol{\Omega}(\mathbf{h}) - \boldsymbol{\Omega}'(\mathbf{h})$,

5. $\sigma_{ij}(\mathbf{h}) = \omega_{ij}(\mathbf{h}) - \omega_{iD}(\mathbf{h}) - \omega_{Dj}(\mathbf{h}) + \omega_{DD}(\mathbf{h})$, *or in matrix notation,*

   $\boldsymbol{\Sigma}(\mathbf{h}) = \mathbf{F}\boldsymbol{\Omega}(\mathbf{h})\mathbf{F}'$,

6. $\xi_{ij}(\mathbf{h}) = \omega_{ij}(\mathbf{h}) - \bar{\omega}_{i\cdot}(\mathbf{h}) - \bar{\omega}_{\cdot j}(\mathbf{h}) + \bar{\bar{\omega}}_{\cdot\cdot}(\mathbf{h})$; *or, in matrix notation,*

   $\boldsymbol{\Xi}(\mathbf{h}) = \mathbf{G}\boldsymbol{\Omega}(\mathbf{h})\mathbf{G}'$,

7. $\sigma_{ij\cdot k\ell}(\mathbf{h}) = \omega_{ij}(\mathbf{h}) - \omega_{i\ell}(\mathbf{h}) - \omega_{kj}(\mathbf{h}) + \omega_{k\ell}(\mathbf{h})$.

Definition 3.13 *The log cross-covariance of any two components of an r-basis with indices i and j is* symmetric *over $\bar{H}$, if it is symmetric in the indices for every $\mathbf{h} \in \bar{H}$:*

$$\omega_{ij}(\mathbf{h}) = \omega_{ji}(\mathbf{h}).$$

Hence, if the log cross-covariances associated with $\mathbf{W}(\mathbf{u})$ are all symmetric, then

$$\boldsymbol{\Omega}(\mathbf{h}) = \boldsymbol{\Omega}'(\mathbf{h}).$$

The proof of the following statements is straightforward:

Property 3.23 *For every lag* $\mathbf{h}$,

$$\begin{aligned} \omega_{ij}(\mathbf{h}) = \omega_{ji}(\mathbf{h}) &\iff \omega_{ij}(\mathbf{h}) = \omega_{ij}(-\mathbf{h}). \\ \boldsymbol{\Omega}(\mathbf{h}) = \boldsymbol{\Omega}'(\mathbf{h}) &\iff \boldsymbol{\Omega}(\mathbf{h}) = \boldsymbol{\Omega}(-\mathbf{h}). \end{aligned}$$

Property 3.24 *If $\boldsymbol{\Omega}(\mathbf{h}) = \boldsymbol{\Omega}'(\mathbf{h})$, $\mathbf{h} \in \bar{H}$, then the spatial covariance structure of the r-composition $\mathbf{Z}(\mathbf{u})$ obtained from $\mathbf{W}(\mathbf{u})$ is symmetric over $\bar{H}$.*

Note that analogous statements can be put forth for the intrinsic case.

Chapter 4

Concepts of null correlation

Concepts of null correlation for r-compositions are discussed in this chapter, following the methodology developed by J. Aitchison for the statistical analysis of compositional data. This will be combined with G. Matheron's theory of regionalized variables. These concepts are to be understood in the sense of *absence de correlation différée* (absence of deferred correlation), as defined by Matheron (1965). Concepts of null correlation are important not only for spatial-structure analysis of r-compositions, but also for simulation of phenomena that can be described by the use of r-compositions.

The intrinsic analogue to the definitions of null correlation in the second-order stationary case is carried out in parallel in this chapter because the relation between them is of special interest.

All of the following concepts depend in general on the length of the vector $\mathbf{h}$ and also on its direction and sign, that is, they can be defined depending on the length, or set of lengths, and the direction, or set of directions, of $\mathbf{h}$, or both. Therefore, as in Chapter 3, statements will be made for $\mathbf{h} \in H$, where H stands for a set of vectors with specified range of directions and range of lengths. Here, for example, H may contain all possible directions and lengths for $\mathbf{h}$, except $\mathbf{h} = 0$; in this case, statements will be valid only in a spatial sense, but not in a standard nonspatial sense. The set H may be different from the set $\bar{H}$ used to define symmetry, although in some cases $H \subset \bar{H}$ will hold; that justifies the notation used.

The terms *shifted independence* and *internal independence* used by Matheron (1971) are not adopted. The concept of *shifted independence* for lag $\mathbf{h}$ coincides with our *absence of cross-correlation* or *null cross-correlation for lag* $\mathbf{h}$. The *internal independence* applies to a nonstochastic understanding of the spatial correlations and is therefore not discussed here.

In the following it is assumed that the r-composition $\mathbf{Z}(\mathbf{u})$ is either second-order lr stationary, or that it satisfies the lr intrinsic hypothesis. As in preceding chapters, proofs analogous to those of the nonregionalized case or that can be derived directly from the definitions are omitted. Most of them can be found in Aitchison (1986).

4.1 Null lr cross-correlation

Definition 4.1 *An r-composition* $\mathbf{Z}(\mathbf{u})$ *has* null lr cross-correlation *over* H *if, for* i, j, k, ℓ *all different and for every* $\mathbf{h} \in H$, $\sigma_{ij\cdot k\ell}(\mathbf{h}) = 0$.

Recall that for $i = k$ and/or $j = \ell$, by property 3.1.5, the cross-covariance of the corresponding logratios is always zero for every lag $\mathbf{h}$, whereas this is not necessarily the case for $k = \ell$ or for $i = j$. Consequently, this definition does not include the absence of autocorrelation for any logratio (case $i = j$ and $k = \ell$) nor the absence of cross-correlation between logratios with the same component as divisor (case $k = \ell$). Furthermore, note that by property 3.1.4, $\sigma_{ii\cdot k\ell}(\mathbf{h}) = \sigma_{k\ell\cdot ii}(\mathbf{h})$, and therefore the case $i = j$ and k, ℓ, i all different need not be considered separately.

Property 4.1 *If* $\mathbf{Z}(\mathbf{u})$ *has null* lr *cross-correlation for* $\mathbf{h} \in H$, *then* $\mathbf{Z}(\mathbf{u})$ *has null* lr *cross-correlation for* $-\mathbf{h}$.

Proof. Using property 3.1.3, the proof is straightforward. □

Property 4.1 could be stated in terms of the set H, that is, with $\mathbf{h}$, H also contains $-\mathbf{h}$; it follows that in the case of null lr cross-correlation, the spatial covariance structure of $\mathbf{Z}(\mathbf{u})$ is symmetric for any i, j, k, ℓ all different and $\mathbf{h} \in H$. Moreover, symmetry is not restricted to the condition that all indices be different; but, the following holds:

Property 4.2 *If* $\mathbf{Z}(\mathbf{u})$ *has null* lr *cross-correlation over* H, *then the spatial covariance structure of* $\mathbf{Z}(\mathbf{u})$ *is symmetric over* H.

Proof. When all four indices are different, the proof is obvious according to the definition of the null lr cross-correlation of $\mathbf{Z}(\mathbf{u})$. For three different indices i, j, k, apply first property 3.1.6 and then property 3.1.4 and the hypothesis of absence of lr cross-correlation in order to prove the following sequence of equalities (taking i, j, k, ℓ all different):

$$\sigma_{ij\cdot kk}(\mathbf{h}) = \sigma_{i\ell\cdot kk}(\mathbf{h}) = \sigma_{j\ell\cdot kk}(\mathbf{h}) = \sigma_{ji\cdot kk}(\mathbf{h}),$$

with $\mathbf{h} \in H$. For two different indices, apply either property 3.1.5 or 3.1.4 to obtain the desired result. □

In view of property 4.2, which implies that the spatial covariance structure is symmetric if the r-composition has null lr cross-correlation for every lag $\mathbf{h}$, it is not surprising that the following statement can be set forth in the same terms (besides the spatial component) as the nonregionalized counterpart, the proof being analogous.

Property 4.3 *If* $\mathbf{Z}(\mathbf{u})$ *has null* lr *cross-correlation over* H, *then there exists*

$$\vec{\alpha}(\mathbf{h}) = (\, \alpha_1(\mathbf{h}) \quad \alpha_2(\mathbf{h}) \quad \cdots \quad \alpha_D(\mathbf{h})\,)',$$

$\mathbf{h} \in H$, *with* $\alpha_k(\mathbf{h}) = \sigma_{ij \cdot kk}(\mathbf{h})$, $i \neq j$; $i, j \neq k$, *and the following statements hold for* $\mathbf{h} \in H$:

1. $\boldsymbol{\Sigma}(\mathbf{h}) = \mathrm{diag}(\vec{\alpha}_{-D}(\mathbf{h})) + \alpha_D(\mathbf{h})\mathbf{J}$,
2. $\mathbf{T}(\mathbf{h}) = \vec{\alpha}(\mathbf{h})\mathbf{j}' + \mathbf{j}\vec{\alpha}'(\mathbf{h}) - 2\mathrm{diag}(\vec{\alpha}(\mathbf{h}))$,
3. $\boldsymbol{\Xi}(\mathbf{h}) = \mathbf{G}\mathrm{diag}(\vec{\alpha}(\mathbf{h}))\mathbf{G}$,

where

$$\mathrm{diag}(\vec{\alpha}_{-D}(\mathbf{h})) = \begin{pmatrix} \alpha_1(\mathbf{h}) & 0 & \cdots & 0 \\ 0 & \alpha_2(\mathbf{h}) & \cdots & 0 \\ \vdots & \vdots & \ddots & \vdots \\ 0 & 0 & \cdots & \alpha_{D-1}(\mathbf{h}) \end{pmatrix}$$

and

$$\mathrm{diag}(\vec{\alpha}(\mathbf{h})) = \begin{pmatrix} \alpha_1(\mathbf{h}) & 0 & \cdots & 0 \\ 0 & \alpha_2(\mathbf{h}) & \cdots & 0 \\ \vdots & \vdots & \ddots & \vdots \\ 0 & 0 & \cdots & \alpha_D(\mathbf{h}) \end{pmatrix}.$$

As a consequence, each of the covariance functions

$$\sigma_{ij \cdot k\ell}(\mathbf{h}), \quad i, j, k, \ell \in \{1, 2, \ldots, D\}, \mathbf{h} \in H,$$

can be expressed in terms of $\vec{\alpha}(\mathbf{h})$. By property 3.1.5 in particular, for three or four equal indices, the corresponding function is always zero, and for two equal indices the only functions that are not null are (i, k, ℓ all different)

$$\sigma_{ii \cdot k\ell}(\mathbf{h}) = \sigma_{k\ell \cdot ii}(\mathbf{h}) = -\sigma_{i\ell \cdot ki}(\mathbf{h}) = -\sigma_{ki \cdot i\ell}(\mathbf{h}) = \alpha_i(\mathbf{h}). \tag{4.1}$$

The equalities hold by property 3.1.4, and

$$\sigma_{ii\cdot kk}(\mathbf{h}) = \alpha_i(\mathbf{h}) + \alpha_k(\mathbf{h}) \tag{4.2}$$

is obtained using properties 3.4.4 and 4.31.

It is important to point out that although the null lr cross-correlation of $\mathbf{Z}(\mathbf{u})$ in H implies that the spatial covariance structure of $\mathbf{Z}(\mathbf{u})$ is symmetric in H, it is not necessarily true that $0 \in H$; so the spatial covariance structure of $\mathbf{Z}(\mathbf{u})$ might be symmetric in a larger set $\bar{H}$ and the condition $\bar{H} \supset H$ will be satisfied. Also, if $\mathbf{Z}(\mathbf{u})$ has null lr cross-correlation for $\mathbf{h} = 0$, it does not imply that it has null lr cross-correlation for $\mathbf{h} \neq 0, \mathbf{h} \in \bar{H}$. Hence, a high correlation in the classic sense can occur with a slight or nonexistent spatial cross-correlation and, conversely, a strong spatial dependence can occur together with a null correlation in the classic sense. For other details, see Matheron (1965).

Another important aspect of null lr cross-correlation is that it is defined for all possible combinations of four different indices. But this definition does not include the cases of absence of cross-correlation in certain directions or for certain distances for one or only some combinations of indices in a predefined manner. In these cases symmetry is not a necessary consequence of absence of correlation, thus absence of correlation and symmetry do not necessarily go together when only a part of the possible logratios are considered. These considerations do not apply to the intrinsic spatial covariance structure, which is symmetric by definition.

Definition 4.2 *An r-composition* $\mathbf{Z}(\mathbf{u})$ *has* null intrinsic lr cross-correlation *over* H*, if for every lag* $\mathbf{h} \in H$*,* $\gamma_{ij\cdot k\ell}(\mathbf{h}) = 0$ *for any* i, j, k, ℓ *all different.*

As before, the intrinsic spatial covariance structure being symmetric, the null intrinsic lr cross-correlation over H of $\mathbf{Z}(\mathbf{u})$ allows a specification for its intrinsic spatial covariance structure through a D-vector $\vec{\beta}(\mathbf{h})$, $\mathbf{h} \in H$, of covariance functions.

Property 4.4 *If* $\mathbf{Z}(\mathbf{u})$ *has null intrinsic* lr *cross-correlation over* H*, then there exists*

$$\vec{\beta}(\mathbf{h}) = (\,\beta_1(\mathbf{h}) \quad \beta_2(\mathbf{h}) \quad \cdots \quad \beta_D(\mathbf{h})\,)',$$

$\mathbf{h} \in H$*, with* $\beta_k(\mathbf{h}) = \gamma_{ij\cdot kk}(\mathbf{h})$*,* $i \neq j;\ i, j \neq k$*, and the following statements hold for* $\mathbf{h} \in H$*:*

1. $\mathbf{\Psi}(\mathbf{h}) = \mathrm{diag}(\vec{\beta}_{-D}(\mathbf{h})) + \beta_D(\mathbf{h})\mathbf{J}$,
2. $\mathbf{\Gamma}(\mathbf{h}) = \vec{\beta}(\mathbf{h})\mathbf{j}' + \mathbf{j}\vec{\beta}'(\mathbf{h}) - 2\mathrm{diag}(\vec{\beta}(\mathbf{h}))$,

3. $\mathbf{\Delta}(\mathbf{h}) = \mathbf{G}\text{diag}(\vec{\beta}(\mathbf{h}))\mathbf{G}$,

where

$$\text{diag}(\vec{\beta}_{-D}(\mathbf{h})) = \begin{pmatrix} \beta_1(\mathbf{h}) & 0 & \cdots & 0 \\ 0 & \beta_2(\mathbf{h}) & \cdots & 0 \\ \vdots & \vdots & \ddots & \vdots \\ 0 & 0 & \cdots & \beta_{D-1}(\mathbf{h}) \end{pmatrix}$$

and

$$\text{diag}(\vec{\beta}(\mathbf{h})) = \begin{pmatrix} \beta_1(\mathbf{h}) & 0 & \cdots & 0 \\ 0 & \beta_2(\mathbf{h}) & \cdots & 0 \\ \vdots & \vdots & \ddots & \vdots \\ 0 & 0 & \cdots & \beta_D(\mathbf{h}) \end{pmatrix}.$$

To express the intrinsic covariance structure in terms of the components of $\vec{\beta}(\mathbf{h})$, properties 3.11.3 and 3.11.4, as well as property 3.13.3, together with property 4.4.1, must be used. The only functions that are not zero in this case are:

$$\gamma_{ii\cdot k\ell}(\mathbf{h}) = \gamma_{k\ell\cdot ii}(\mathbf{h}) = -\gamma_{i\ell\cdot ki}(\mathbf{h}) = -\gamma_{ki\cdot i\ell}(\mathbf{h}) = \beta_i(\mathbf{h}) \qquad (4.3)$$

and

$$\gamma_{ii\cdot kk}(\mathbf{h}) = \beta_i(\mathbf{h}) + \beta_k(\mathbf{h}), \qquad (4.4)$$

for i, k, ℓ all different.

4.2 Null lr autocorrelation

The following definition of absence of correlation is of interest for describing the spatial behavior of an individual logratio.

Definition 4.3 *An r-composition* $\mathbf{Z}(\mathbf{u})$ *has* null lr autocorrelation *over* H *if, for every lag* $\mathbf{h} \in H$*, and for all* $i, j \in \{1, 2, \ldots, D\}$*,* $\tau_{i\cdot j}(\mathbf{h}) = 0$.

Recall that by property 3.3.1, $\tau_{i\cdot i}(\mathbf{h}) = 0$ for every lag $\mathbf{h}$; therefore it is not necessary to exclude here the case $i = j$. By the same property, it holds that if an r-composition $\mathbf{Z}(\mathbf{u})$ has null lr autocorrelation over H, then $\mathbf{T}(\mathbf{h}) = \mathbf{0}$ for every $\mathbf{h} \in H$, and the converse holds. Furthermore, in connection with the absence of lr autocorrelation, using properties 3.3.3 and 3.3.4, the following statements can be set forth:

Property 4.5 *If* $\tau_{n\cdot m}(\mathbf{h}) = 0$ *for all* $n, m \in \{i, j, k, \ell\}$*, then* $\sigma_{ij\cdot k\ell}(\mathbf{h}) + \sigma_{ji\cdot \ell k}(\mathbf{h}) = 0$.

Property 4.6 *If $\tau_{n \cdot m}(\mathbf{h}) = 0$ for all $n, m \in \{i, j, k, \ell\}$ and if the spatial covariance structure is symmetric for every lag $\mathbf{h} \in H$, then $\sigma_{ij \cdot k\ell}(\mathbf{h}) = 0$, $\mathbf{h} \in H$.*

A consequence of property 4.6 is that only in the case where $\mathbf{Z}(\mathbf{u})$ has null lr autocorrelation over H, *and* the spatial covariance structure is symmetric, does $\mathbf{Z}(\mathbf{u})$ have null lr cross-correlation over H. But in this case, resulting absence of correlation is more restrictive because, as shall be seen later, it implies absence of any type of spatial structure.

Here no intrinsic counterpart exists because the existence of a variance is implicit in the lr semivariogram. Nevertheless, by property 3.16.2, the following holds:

Property 4.7 *For any $\mathbf{h}_0 \neq 0$, $\mathbf{T}(h_0) = \mathbf{0} \iff \mathbf{\Gamma}(h_0) = \mathbf{T}(0)$.*

Thus, if there is no lr autocorrelation for all $\mathbf{h} \neq 0$, then all lr semivariograms are constant, and the converse is true.

The following properties are obtained using properties 4.5, 4.6, 3.8.1, 3.8.2, 3.4.3, and 3.6.3, as well as the corresponding properties in the case of symmetry:

Property 4.8 *If $\mathbf{Z}(\mathbf{u})$ has null* lr *autocorrelation over H then, for all $\mathbf{h} \in H$,*

1. $\sigma_{ij \cdot k\ell}(\mathbf{h}) = -\sigma_{ij \cdot k\ell}(-\mathbf{h})$, $\forall i, j, k, \ell$,

2. $\mathbf{\Sigma}(\mathbf{h}) = -\mathbf{\Sigma}(-\mathbf{h})$,

3. $\mathbf{\Xi}(\mathbf{h}) = -\mathbf{\Xi}(-\mathbf{h})$.

Property 4.9 *$\mathbf{Z}(\mathbf{u})$ has null* lr *autocorrelation and spatial covariance structure symmetric over H if and only if one of the following conditions is satisfied:*

1. $\sigma_{ij \cdot k\ell}(\mathbf{h}) = 0$, $\forall \mathbf{h} \in H$, $\forall i, j, k, \ell$,

2. $\mathbf{\Sigma}(\mathbf{h}) = \mathbf{0}$, $\forall \mathbf{h} \in H$,

3. $\mathbf{\Xi}(\mathbf{h}) = \mathbf{0}$, $\forall \mathbf{h} \in H$.

Note that property 4.9.3 implies that the vector $\vec{\alpha}(\mathbf{h})$ of property 4.3 is identically zero for all $\mathbf{h} \in H$, that is, there is no spatial structure over H.

4.3 Null alr cross-correlation

Definition 4.4 *An r-composition* $\mathbf{Z}(\mathbf{u})$ *has* null alr cross-correlation *over* H *if for every lag* $\mathbf{h} \in H$ *and for all* $i, j \in \{1, 2, \ldots, D\}$, $i \neq j$, $\sigma_{ij}(\mathbf{h}) = 0$.

Null alr cross-correlation corresponds to absence of correlation between alr transformed variables, but admits the existence of autocorrelation for each alr transformed variable involved. This type of absence of correlation is of special interest for cokriging because, as can easily be deduced from the following property, in the case of null alr cross-correlation for every $\mathbf{h}$, ordinary cokriging and ordinary kriging of the alr transformed variables are equivalent.

Property 4.10 *If an r-composition* $\mathbf{Z}(\mathbf{u})$ *has null* alr *cross-correlation over* H, *then the* alr *cross-covariance matrix is diagonal:* $\boldsymbol{\Sigma}(\mathbf{h}) = \text{diag}(\boldsymbol{\Sigma}(\mathbf{h}))$.

Property 4.11 *If an r-composition* $\mathbf{Z}(\mathbf{u})$ *has null* alr *cross-correlation over* H, *then* $\mathbf{Z}(\mathbf{u})$ *has null* lr *cross-correlation over* H.

Note that the converse is not true, as can be seen from property 4.3. There it is stated that in the case of absence of lr cross-correlation, the cross-covariance matrix $\boldsymbol{\Sigma}(\mathbf{h}) = \text{diag}(\vec{\alpha}_{-D}(\mathbf{h})) + \alpha_D(\mathbf{h})\mathbf{J}$, while in the case of absence of alr cross-correlation over H, $\alpha_D(\mathbf{h}) = 0$ and $\alpha_k(\mathbf{h}) = \sigma_{kk}(\mathbf{h})$ for all $k \neq D$, and for all $\mathbf{h} \in H$.

Here again equivalent statements can be made for the intrinsic case. The statements are analogous to those of the nonregionalized case due to the symmetry of the intrinsic spatial covariance structure.

Definition 4.5 *An r-composition* $\mathbf{Z}(\mathbf{u})$ *has* null intrinsic alr cross-correlation *over* H *if for every lag* $\mathbf{h} \in H$ *and for all* $i, j \in \{1, 2, \ldots, D\}$, $i \neq j$, $\psi_{ij}(\mathbf{h}) = 0$.

Property 4.12 *If an r-composition* $\mathbf{Z}(\mathbf{u})$ *has null intrinsic* alr *cross-correlation over* H, *then the intrinsic* alr *cross-covariance matrix is a diagonal matrix:* $\boldsymbol{\Psi}(\mathbf{h}) = \text{diag}(\boldsymbol{\Psi}(\mathbf{h}))$.

Property 4.13 *If an r-composition* $\mathbf{Z}(\mathbf{u})$ *has null intrinsic* alr *cross-correlation over* H, *then* $\mathbf{Z}(\mathbf{u})$ *has null intrinsic* lr *cross-correlation over* H.

The converse is not true, as can be deduced from property 4.4.

4.4 Null clr cross-correlation

As in the nonregionalized case, null clr cross-correlation can be defined:

Definition 4.6 *An r-composition* $\mathbf{Z}(\mathbf{u})$ *has* null clr cross-correlation *over* H *if for every lag* $\mathbf{h} \in H$ *and* $i, j \in \{1, 2, \ldots, D\}$, $i \neq j$, $\xi_{ij}(\mathbf{h}) = 0$.

The clr cross-covariance matrix being singular by property 3.6.5, it holds that if an r-composition has null clr cross-correlation, then $\boldsymbol{\Xi}(\mathbf{h}) = \mathbf{0}$.

Property 4.14 *If an r-composition* $\mathbf{Z}(\mathbf{u})$ *has null* clr *cross-correlation over* H *then the following statements hold:*

1. $\mathbf{Z}(\mathbf{u})$ *has null* lr *cross-correlation over* H,
2. $\boldsymbol{\Sigma}(\mathbf{h}) = \mathbf{0}$, $\forall \mathbf{h} \in H$,
3. $\mathbf{T}(\mathbf{h}) = \mathbf{0}$, $\forall \mathbf{h} \in H$.

This property results directly from property 3.6.4 and properties 3.8.5 and 3.8.6. Accordingly, absence of clr cross-correlation is the strongest type of absence of cross-correlation, in particular because, besides null lr cross-correlation, the stated property implies both null alr cross-correlation and null lr autocorrelation over H. The converse is not true in general, as can be deduced from properties 3.8.2 and 3.8.4.

Using properties 4.9 and 4.14, the following statement can be proven:

Property 4.15 *If the spatial covariance structure of an r-composition* $\mathbf{Z}(\mathbf{u})$ *is symmetric, then* $\mathbf{Z}(\mathbf{u})$ *has null* lr *autocorrelation if and only if* $\mathbf{Z}(\mathbf{u})$ *has null* clr *cross-correlation.*

Similar relationships hold in the intrinsic case:

Definition 4.7 *An r-composition* $\mathbf{Z}(\mathbf{u})$ *has* null intrinsic clr cross-correlation *over* H *if for every lag* $\mathbf{h} \in H$ *and for all* $i, j \in \{1, 2, \ldots, D\}$, $i \neq j$, $\delta_{ij}(\mathbf{h}) = 0$.

As before, due to the singularity of the intrinsic clr cross-covariance matrix, in the case of null intrinsic clr cross-correlation $\boldsymbol{\Delta}(\mathbf{h}) = \mathbf{0}$; also, absence of intrinsic clr cross-correlation implies null intrinsic lr cross-correlation and null intrinsic alr cross-correlation.

4.5 Composition invariance

Definition 4.8 *An r-basis* $\mathbf{W}(\mathbf{u})$ *with associated r-composition* $\mathbf{Z}(\mathbf{u}) = \check{C}(\mathbf{W}(\mathbf{u}))$ *and size* $t(\mathbf{u}+\mathbf{h}) = \mathbf{W}(\mathbf{u}+\mathbf{h})'\mathbf{j}$ *is said to be*

1. composition-invariant *over* H *if, for all* $\mathbf{h} \in H$, $\mathbf{Z}(\mathbf{u})$ *is statistically independent of* $t(\mathbf{u}+\mathbf{h})$,
2. intrinsic composition-invariant *over* H *if, for all* $\mathbf{h} \in H$, $Z(\mathbf{u}) - Z(\mathbf{u}+\mathbf{h})$ *is statistically independent of* $t(\mathbf{u}) - t(\mathbf{u}+\mathbf{h})$.

Examples include:

1. the question about invariance of composition of geochemical components versus thickness—in this case the corresponding basis is the accumulation of the individual components,
2. the relationship of total thickness to seam thicknesses in coal deposits.

A way to analyze composition invariance in the spatial context is to check cross-covariances of the alr transformed composition with the log-transformed size.

Definition 4.9 *An r-basis has* null log cross-correlation *over* H *if, for all* $\mathbf{h} \in H$ *and* $i, j \in \{1, 2, \ldots, D\}$, i, j *different,* $\omega_{ij}(\mathbf{h}) = 0$.

Let $\omega(\mathbf{h}) = (\omega_{11}(\mathbf{h}) \;\; \omega_{22}(\mathbf{h}) \;\; \cdots \;\; \omega_{DD}(\mathbf{h}))'$, and $\omega_{-D}(\mathbf{h})$ be the same vector without the last component.

Property 4.16 *If* $\mathbf{W}(\mathbf{u})$ *has null log cross-correlation over* H*, then, for all* $\mathbf{h} \in H$,

1. $\mathbf{\Sigma}(\mathbf{h}) = \mathrm{diag}(\omega_{-D}(\mathbf{h})) + \omega_D \mathbf{J}$,
2. $\mathbf{\Xi}(\mathbf{h}) = \mathbf{G}\mathrm{diag}(\omega(\mathbf{h}))\mathbf{G}$,
3. $\mathbf{T}(\mathbf{h}) = \omega(\mathbf{h})\mathbf{j}' + \mathbf{j}\omega'(\mathbf{h}) - 2\mathrm{diag}(\omega(\mathbf{h}))$,
4. $\mathbf{Z}(\mathbf{u})$ *has null* lr *cross-correlation.*

A consequence of this property is that if $\mathbf{W}(\mathbf{u})$ has null log cross-correlation over H, the vector $\vec{\alpha}(\mathbf{h})$ of property 4.3 is identical to the vector $\omega(\mathbf{h})$.

Analogous statements for the intrinsic case can be set forth easily, but are omitted here as they are seldom required and, at the same time, are straightforward.

4.6 Relationship between concepts of null cross-correlation

The results described in previous sections which relate the different concepts of null cross-correlation can be presented in the form of diagrams. These depend on whether or not the hypothesis of symmetry of the spatial covariance structure is assumed. Here only the diagrams corresponding to the hypothesis of second-order stationarity are displayed, but analogous diagrams can be developed for the intrinsic case. Figure 4.1 represents relationships when no symmetry can be assumed, while Figure 4.2 represents corresponding relationships in presence of symmetry. The numbers beside the arrows refer to the properties where the implications are stated. It is assumed that the definitions of absence of cross-correlation are satisfied over a set H, and that the indices involved are all different.

null log cross-correlation $(\omega_{ij}(\mathbf{h}) = 0)$		
$\Downarrow$ 4.16.4		
null lr cross-correlation $(\sigma_{ij \cdot k\ell}(\mathbf{h}) = 0)$	$\overset{4.11}{\Longleftarrow}$	null alr cross-correlation $(\sigma_{ij}(\mathbf{h}) = 0)$
		$\Uparrow$ 4.14.2
null lr autocorrelation $(\tau_{i \cdot j}(\mathbf{h}) = 0)$	$\overset{4.14.3}{\Longleftarrow}$	null clr cross-correlation $(\xi_{ij}(\mathbf{h}) = 0)$

Figure 4.1: Relationships between concepts of null correlation assuming absence of symmetry in the spatial covariance structure.

null log cross-correlation $(\omega_{ij}(\mathbf{h}) = 0)$		
$\Downarrow$ 4.16.4		
null lr cross-correlation $(\sigma_{ij \cdot k\ell}(\mathbf{h}) = 0)$	$\overset{4.11}{\Longleftrightarrow}$	null alr cross-correlation $(\sigma_{ij}(\mathbf{h}) = 0)$
$\Updownarrow$ 4.9.1		$\Updownarrow$ 4.9.2
null lr autocorrelation $(\tau_{i \cdot j}(\mathbf{h}) = 0)$	$\overset{4.9.3}{\Longleftrightarrow}$	null clr cross-correlation $(\xi_{ij}(\mathbf{h}) = 0)$

Figure 4.2: Relationships between concepts of null correlation assuming symmetry in the spatial covariance structure.

The diagrams shown in Figures 4.1 and 4.2 make clear the fact that when the spatial covariance structure is not symmetric, the absence of clr cross-covariance is the strongest type of null cross-correlation; while in the case of symmetry, null clr cross-covariance and null lr autocovariance are equivalent.

4.7 Subcompositional invariance and partition independence

The concepts of subcompositional invariance and partition independence as stated for nonregionalized compositions form an important tool for analysis of phenomena characterized by means of a composition. This is true because interest frequently lies in the relationship between groups of variables such as major and minor elements—more so than in the relationship between individual variables. The extension of these concepts to regionalized compositions, like the other concepts of absence of correlation discussed in preceding sections, is straightforward. These concepts may provide a means of determining whether subcompositions have in some sense an independent evolution in space and/or time. The essential theoretical difference between the regionalized and the nonregionalized case is, again, the possible nonsymmetry of the spatial covariance structure; therefore, the presentation of involved concepts is held very concise. Difficulties arise in practice due to the fact that only a partial realization of an r-composition is usually available. Aitchison (1986, chapter 10) gives an extensive discussion of the nonregionalized case, which can be extended directly to r-compositions.

Consider a partition of order one (see definition 2.9) of $\mathbf{Z}(\mathbf{u})$, $\mathbf{u} \in \Omega$,

$$P_1(\mathbf{Z}(\mathbf{u})) = (\mathbf{s}_1(\mathbf{u}), \mathbf{s}_2(\mathbf{u}); \mathbf{V}(\mathbf{u})),$$

where

$$\mathbf{V}(\mathbf{u}) = (\, t_1(\mathbf{u}) \quad t_2(\mathbf{u})\,)', \text{ with } t_1(\mathbf{u}) = \sum_{i=1}^{m} Z_i(\mathbf{u}),\ t_2(\mathbf{u}) = \sum_{j=m+1}^{D} Z_j(\mathbf{u}),$$

$$\mathbf{s}_1(\mathbf{u}) = \check{C}\,(\, Z_1(\mathbf{u}) \quad Z_2(\mathbf{u}) \quad \cdots \quad Z_m(\mathbf{u})\,)',$$

$$\mathbf{s}_2(\mathbf{u}) = \check{C}\,(\, Z_{m+1}(\mathbf{u}) \quad Z_{m+2}(\mathbf{u}) \quad \cdots \quad Z_D(\mathbf{u})\,)'.$$

Take

$$\mathbf{Y}_1(\mathbf{u}) = \mathrm{alr}(\mathbf{s}_1(\mathbf{u})),\ \mathbf{Y}_2(\mathbf{u}) = \mathrm{alr}(\mathbf{s}_2(\mathbf{u})),$$

$$\mathbf{U}(\mathbf{u}) = \mathrm{alr}(\mathbf{V}(\mathbf{u})) = \ln(t_1(\mathbf{u})/t_2(\mathbf{u})),$$

$$\mathbf{p}_1^*(\mathbf{u}) = (\, \mathbf{Y}_1(\mathbf{u}) \quad \mathbf{Y}_2(\mathbf{u}) \quad \mathbf{U}(\mathbf{u})\,)'.$$

The spatial dependencies between the subcompositions $\mathbf{s}_1(\mathbf{u})$, $\mathbf{s}_2(\mathbf{u})$, and the r-amalgamation $\mathbf{V}(\mathbf{u})$ are reflected in the matrix-valued covariance function

$$\mathrm{Cov}[\mathbf{p}_1^*(\mathbf{u}), \mathbf{p}_1^*(\mathbf{u}+\mathbf{h})] = \mathbf{\Sigma}^*(\mathbf{h}) = \begin{pmatrix} \mathbf{\Sigma}_{11}(\mathbf{h}) & \mathbf{\Sigma}_{12}(\mathbf{h}) & \mathbf{\Sigma}_{1u}(\mathbf{h}) \\ \mathbf{\Sigma}_{21}(\mathbf{h}) & \mathbf{\Sigma}_{22}(\mathbf{h}) & \mathbf{\Sigma}_{2u}(\mathbf{h}) \\ \mathbf{\Sigma}_{u1}(\mathbf{h}) & \mathbf{\Sigma}_{u2}(\mathbf{h}) & \mathbf{\Sigma}_{uu}(\mathbf{h}) \end{pmatrix}, \qquad (4.5)$$

where

$$\begin{aligned} \mathbf{\Sigma}_{ij}(\mathbf{h}) &= \mathrm{Cov}\left[\mathbf{Y}_i(\mathbf{u}), \mathbf{Y}_j(\mathbf{u}+\mathbf{h})\right], \quad i,j = 1,2; \\ \mathbf{\Sigma}_{ui}(\mathbf{h}) &= \mathrm{Cov}\left[\mathbf{U}(\mathbf{u}), \mathbf{Y}_i(\mathbf{u}+\mathbf{h})\right], \quad i = 1,2; \\ \mathbf{\Sigma}_{iu}(\mathbf{h}) &= \mathrm{Cov}\left[\mathbf{Y}_i(\mathbf{u}), \mathbf{U}(\mathbf{u}+\mathbf{h})\right], \quad i = 1,2; \\ \mathbf{\Sigma}_{uu}(\mathbf{h}) &= \mathrm{Cov}\left[\mathbf{U}(\mathbf{u}), \mathbf{U}(\mathbf{u}+\mathbf{h})\right]. \end{aligned}$$

Note that, for example, $\mathbf{\Sigma}_{12}(\mathbf{h})$ is a matrix of cross-covariance functions with dimension $m \times (D-m)$, whereas $\mathbf{\Sigma}_{uu}(\mathbf{h})$ is just one single autocovariance function, and $\mathbf{\Sigma}_{u2}(\mathbf{h})$ is a row-vector of cross-covariances. In general, none of the above-mentioned matrices will have the same dimension, so care must be taken. Accordingly when any of these matrices are assumed to be equal to $\mathbf{0}$, it is understood that the matrix $\mathbf{0}$ is a matrix with all its elements equal to zero and the appropriate dimension.

The fact that the spatial covariance structure of $\mathbf{Z}(\mathbf{u})$ is not necessarily symmetric implies that, in general, $\mathbf{\Sigma}^*(\mathbf{h}) \neq (\mathbf{\Sigma}^*(\mathbf{h}))'$ and $\mathbf{\Sigma}^*(\mathbf{h}) \neq \mathbf{\Sigma}^*(-\mathbf{h})$; at the same time it implies that $\mathbf{\Sigma}^*(\mathbf{h}) = (\mathbf{\Sigma}^*(-\mathbf{h}))'$ at all times. Hence, the following concepts of subcompositional invariance and partition independence can be set forth, although it is possible to consider similar hypotheses without an implicit statement of symmetry.

Definition 4.10 *Subcompositional invariance and partition independence.*

1. $\mathbf{p}_1^*(\mathbf{u})$ *has first spatial subcompositional invariance over* H *if* $\mathbf{Y}_1(\mathbf{u})$ *and* $\mathbf{U}(\mathbf{u}+\mathbf{h})$, *as well as* $\mathbf{Y}_1(\mathbf{u}+\mathbf{h})$ *and* $\mathbf{U}(\mathbf{u})$, *are not correlated for all* $\mathbf{h} \in H$; i.e., *if for all* $\mathbf{h} \in H$, $\mathbf{\Sigma}_{1u}(\mathbf{h}) = \mathbf{0}$ *and* $\mathbf{\Sigma}_{u1}(\mathbf{h}) = \mathbf{0}$, *or equivalently, if for all* $\mathbf{h} \in H$, $\mathbf{\Sigma}_{1u}(-\mathbf{h}) = \mathbf{0}$ *and* $\mathbf{\Sigma}_{u1}(-\mathbf{h}) = \mathbf{0}$. *Substituting into equation 4.5, the expression*

$$\mathbf{\Sigma}^*(\mathbf{h}) = \begin{pmatrix} \mathbf{\Sigma}_{11}(\mathbf{h}) & \mathbf{\Sigma}_{12}(\mathbf{h}) & \mathbf{0} \\ \mathbf{\Sigma}_{21}(\mathbf{h}) & \mathbf{\Sigma}_{22}(\mathbf{h}) & \mathbf{\Sigma}_{2u}(\mathbf{h}) \\ \mathbf{0} & \mathbf{\Sigma}_{u2}(\mathbf{h}) & \mathbf{\Sigma}_{uu}(\mathbf{h}) \end{pmatrix}$$

is obtained for all $\mathbf{h} \in H$.

2. $\mathbf{p}_1^*(\mathbf{u})$ *has second spatial subcompositional invariance over* H *if, for all* $\mathbf{h} \in H$,

$$\boldsymbol{\Sigma}^*(\mathbf{h}) = \begin{pmatrix} \boldsymbol{\Sigma}_{11}(\mathbf{h}) & \boldsymbol{\Sigma}_{12}(\mathbf{h}) & \boldsymbol{\Sigma}_{1u}(\mathbf{h}) \\ \boldsymbol{\Sigma}_{21}(\mathbf{h}) & \boldsymbol{\Sigma}_{22}(\mathbf{h}) & \mathbf{0} \\ \boldsymbol{\Sigma}_{u1}(\mathbf{h}) & \mathbf{0} & \boldsymbol{\Sigma}_{uu}(\mathbf{h}) \end{pmatrix};$$

3. $\mathbf{p}_1^*(\mathbf{u})$ *has spatial subcompositional invariance over* H *if, for all* $\mathbf{h} \in H$,

$$\boldsymbol{\Sigma}^*(\mathbf{h}) = \begin{pmatrix} \boldsymbol{\Sigma}_{11}(\mathbf{h}) & \mathbf{0} & \boldsymbol{\Sigma}_{1u}(\mathbf{h}) \\ \mathbf{0} & \boldsymbol{\Sigma}_{22}(\mathbf{h}) & \boldsymbol{\Sigma}_{2u}(\mathbf{h}) \\ \boldsymbol{\Sigma}_{u1}(\mathbf{h}) & \boldsymbol{\Sigma}_{u2}(\mathbf{h}) & \boldsymbol{\Sigma}_{uu}(\mathbf{h}) \end{pmatrix};$$

4. $\mathbf{p}_1^*(\mathbf{u})$ *has spatial partition independence over* H *if, for all* $\mathbf{h} \in H$,

$$\boldsymbol{\Sigma}^*(\mathbf{h}) = \begin{pmatrix} \boldsymbol{\Sigma}_{11}(\mathbf{h}) & \mathbf{0} & \mathbf{0} \\ \mathbf{0} & \boldsymbol{\Sigma}_{22}(\mathbf{h}) & \mathbf{0} \\ \mathbf{0} & \mathbf{0} & \boldsymbol{\Sigma}_{uu}(\mathbf{h}) \end{pmatrix};$$

5. $\mathbf{p}_1^*(\mathbf{u})$ *has spatial subcompositional invariance over* H *if, for all* $\mathbf{h} \in H$,

$$\boldsymbol{\Sigma}^*(\mathbf{h}) = \begin{pmatrix} \boldsymbol{\Sigma}_{11}(\mathbf{h}) & \boldsymbol{\Sigma}_{12}(\mathbf{h}) & \mathbf{0} \\ \boldsymbol{\Sigma}_{21}(\mathbf{h}) & \boldsymbol{\Sigma}_{22}(\mathbf{h}) & \mathbf{0} \\ \mathbf{0} & \mathbf{0} & \boldsymbol{\Sigma}_{uu}(\mathbf{h}) \end{pmatrix};$$

6. $\mathbf{p}_1^*(\mathbf{u})$ *has spatial neutrality on the right over* H *if, for all* $\mathbf{h} \in H$,

$$\boldsymbol{\Sigma}^*(\mathbf{h}) = \begin{pmatrix} \boldsymbol{\Sigma}_{11}(\mathbf{h}) & \mathbf{0} & \boldsymbol{\Sigma}_{1u}(\mathbf{h}) \\ \mathbf{0} & \boldsymbol{\Sigma}_{22}(\mathbf{h}) & \mathbf{0} \\ \boldsymbol{\Sigma}_{u1}(\mathbf{h}) & \mathbf{0} & \boldsymbol{\Sigma}_{uu}(\mathbf{h}) \end{pmatrix};$$

7. $\mathbf{p}_1^*(\mathbf{u})$ *has spatial neutrality on the left over* H *if, for all* $\mathbf{h} \in H$,

$$\boldsymbol{\Sigma}^*(\mathbf{h}) = \begin{pmatrix} \boldsymbol{\Sigma}_{11}(\mathbf{h}) & \mathbf{0} & \mathbf{0} \\ \mathbf{0} & \boldsymbol{\Sigma}_{22}(\mathbf{h}) & \boldsymbol{\Sigma}_{2u}(\mathbf{h}) \\ \mathbf{0} & \boldsymbol{\Sigma}_{u2}(\mathbf{h}) & \boldsymbol{\Sigma}_{uu}(\mathbf{h}) \end{pmatrix}.$$

For these concepts of subcompositional invariance and partition independence the following properties can be set forth:

Property 4.17 *The spatial partition independence over* H *given in definition 4.10.4 implies spatial subcompositional invariance over* H*, as defined in item 5, spatial neutrality on the right over* H*, as defined in item 6, and spatial neutrality on the left over* H*, as defined in item 7.*

Property 4.18 *Taken two at a time, concepts 5, 6, and 7 of definition 4.10 over H imply spatial partition independence over H, as defined in item 4 of definition 4.10.*

From this, as stated by Aitchison (1986, chapter 10), for the nonregionalized case, a lattice structure results. This lattice structure should alleviate the burden of decision-making when the following criterion is applied. If a simple hypothesis can reasonably be accepted, it is not to be replaced by a more complex one, even if the latter seems to explain more (principle of parsimony). Analogous hypotheses of independence can be set up for partitions of higher order.

Subcompositional invariance and partition independence can be illustrated by the example of the breaking down of a measurement vector into main and secondary components. If $\mathbf{s}_1(\mathbf{u})$ stands for the r-subcomposition of the main components and $\mathbf{s}_2(\mathbf{u})$ stands for the r-subcomposition of the secondary components, then $t_1(\mathbf{u})$ is the total portion of the main components and $t_2(\mathbf{u})$ is the total portion of the secondary components in each specimen. The different hypotheses of invariance and independence are aimed then at characterizing the spatial interaction between main and secondary components and the relative proportion of these r-subcompositions in each specimen.

Chapter 5

Cokriging

The problem of estimation of a coregionalization of size q using cokriging will be discussed in this chapter. Cokriging—a multivariate extension of kriging—is the usual procedure applied to multivariate regionalized problems within the framework of geostatistics. Its fundament is a distribution-free, linear, unbiased estimator with minimum estimation variance, although the absence of constraints on the estimator is an implicit assumption that the multidimensional real space is the sample space of the variables under consideration.

If a multivariate normal distribution can be assumed for the vector random function, then the simple kriging estimator is identical with the conditional expectation, given a sample of size N. See Journel (1977, pp. 576–577), Journel (1980, pp. 288–290), Cressie (1991, p. 110), and Diggle, Tawn, and Moyeed (1998, p. 300) for further details. This estimator is in general the best possible linear estimator, as it is unbiased and has minimum estimation variance, but it is not very robust in the face of strong departures from normality. Therefore, for the estimation of regionalized compositions other distributions must also be taken into consideration.

Recall that compositions cannot follow a multivariate normal distribution by definition, their sample space being the simplex. Consequently, regionalized compositions in general cannot be modeled under explicit or implicit assumptions of multivariate Gaussian processes. Here only the multivariate lognormal and additive logistic normal distributions will be addressed. Besides the logarithmic and additive logratio transformations, others can be applied, such as the multivariate Box-Cox transformation, as stated by Andrews et al. (1971), Rayens and Srinivasan (1991), and Barceló-Vidal (1996). Furthermore, distributions such as the multiplicative logistic normal distribution introduced by Aitchison (1986, p. 131) or the additive logistic skew-normal distribution defined by Azzalini and Dalla Valle (1996) can be investigated in a similar fashion.

References to the literature for the fundamental principles of the theory discussed in this chapter were given in Chapter 2. Among those, special attention is drawn to the work of Myers (1982), where matrix formulation of cokriging was first presented and the properties included in the first section of this chapter were stated. Matrix formulation shows that there is no significant increase in computing time when several or all variables are estimated, while the amount of work remains essentially the same, as no additional semivariograms or cross-semivariograms need to be computed. For descriptions in the univariate case that are dependent upon the assumed distribution, particularly for the lognormal distribution, refer to the publications of Krige (1978), Journel (1980), Dowd (1982), and Rivoirard (1990); Rendu (1979) treats lognormal kriging and block kriging. As stated by Dowd (1982), lognormal block kriging requires either the assumption of conservation of lognormality from quasi-point supports to blocks, or an iterative solution to the kriging equations. Dowd (1982) states further that, although frequently observed in practice, the assumption of lognormality is statistically invalid, as the probability distribution of a linear combination of lognormal variates is not lognormal. This hypothesis is not necessary when considering estimation of regionalized variables on quasi-point supports using observations with the same support size, and this is the case treated here.

The necessity of assuming quasi-point supports revealed by Dowd (1982) is due to the fact that some regionalized variables and coregionalizations based on a strict point support—such as grade or composition measurements—can be neither normal nor lognormal, as a grade defined on a point support can either be present if the point lies inside a mineralized grain, or absent if it lies inside a grain of waste. As our main interest lies in regionalized compositions and related variables such as regionalized bases and alr transformed coregionalizations, the same reasoning can be applied for the corresponding parent distributions. In the following, therefore, support size is assumed to be small enough that standard formulas for geostatistical analysis devised for point supports can be applied, and large enough that distributional assumptions are reasonable, that is, the support of the variables under study is assumed to be quasi-point.

Any form of kriging is a special case of a corresponding type of cokriging for the particular case of a singular variable.

5.1 The general case of cokriging

Let $\mathbf{X}(\mathbf{u})$, $\mathbf{u} \in \Omega$, be a vector random function, with sample space $\mathbb{R}^q$, defined over a region $\Omega \subset \mathbb{R}^n$ and consider a partial realization, or sample of size N, of the same:

$$\mathbf{X}(\mathbf{u}_k),\ k = 1, 2, \ldots, N,\ \mathbf{u}_k \in \Omega \subset \mathbb{R}^n,\ \mathbf{X}(\mathbf{u}_k) \in \mathbb{R}^q.$$

Assume $\mathbf{X}(\mathbf{u})$ is second-order stationary, i.e., for all $\mathbf{u} \in \Omega$ and every lag $\mathbf{h}$,

1. $\mathrm{E}[\mathbf{X}(\mathbf{u})] = \vec{\mu}$;
2. $\mathrm{Cov}[\mathbf{X}(\mathbf{u}), \mathbf{X}(\mathbf{u} + \mathbf{h})] = \mathbf{\Sigma}(\mathbf{h})$,

where $\vec{\mu}$ does not depend on $\mathbf{u}$ and $\mathbf{\Sigma}(\mathbf{h})$ depends only on $\mathbf{h}$. No symmetry is assumed for the spatial covariances, so in general $\mathbf{\Sigma}(\mathbf{h}) \neq \mathbf{\Sigma}'(\mathbf{h})$ will hold, but recall that at all times $\mathbf{\Sigma}(\mathbf{h}) = \mathbf{\Sigma}'(-\mathbf{h})$.

At location $\mathbf{u}_0$ the unknown $\mathbf{X}(\mathbf{u}_0)$ is estimated from the existing sample by means of a *linear* function:

$$\mathbf{X}^*(\mathbf{u}_0) = \mathbf{c} + \sum_{k=1}^{N} \mathbf{B}_k \mathbf{X}(\mathbf{u}_k), \tag{5.1}$$

where $\mathbf{c} \in \mathbb{R}^q$ is a vector of constants and, for each $k = 1, 2, \ldots, N$, $\mathbf{B}_k$ is a $q \times q$ matrix of weights. Element b_{ij}^k of matrix $\mathbf{B}_k$ measures the influence of the jth component of observation $\mathbf{X}(\mathbf{u}_k)$ on the ith component of $\mathbf{X}^*(\mathbf{u}_0)$. As usual in statistics, $\mathbf{X}(\mathbf{u}_k)$, $k = 1, 2, \ldots, N$, are considered to be random vectors before the actual sample is taken, and vectors of real numbers afterward.

In practice the entire sample is not used for estimation purposes, only the observations located within a given neighborhood that depends on the sampling density and, to a lesser extent, on the range of influence of the variables considered. The neighborhood is defined by the user. As this distinction has no fundamental effect on the equations obtained, it is not considered specifically. The sample size N must be understood as the size of the sample used for estimation at point $\mathbf{u}_0$.

For the proofs in this chapter, note that equation (5.1) can be written in the following form:

$$\mathbf{X}^*(\mathbf{u}_0) = \mathbf{c} + \mathbf{B}\mathbf{X}, \tag{5.2}$$

where

$$\mathbf{B} = \begin{pmatrix} \mathbf{B}_1 & \mathbf{B}_2 & \cdots & \mathbf{B}_N \end{pmatrix}$$

is a $q \times qN$ matrix, and

$$\begin{aligned}\mathbf{X} &= (X_1(\mathbf{u}_1) \quad \cdots \quad X_q(\mathbf{u}_1) \quad \cdots \quad X_1(\mathbf{u}_N) \quad \cdots \quad X_q(\mathbf{u}_N))' \\ &= (\mathbf{X}'(\mathbf{u}_1) \quad \mathbf{X}'(\mathbf{u}_2) \quad \cdots \quad \mathbf{X}'(\mathbf{u}_N))'\end{aligned}$$

is a vector with qN components.

Property 5.1 *For the true but unknown* $\mathbf{X}(\mathbf{u}_0)$ *and for the estimator* $\mathbf{X}^*(\mathbf{u}_0)$ *defined in equation (5.1), the following properties hold:*

1. *the vector of expected values of* $\mathbf{X}^*(\mathbf{u}_0)$ *is* $\mathrm{E}[\mathbf{X}^*(\mathbf{u}_0)] = \mathbf{c} + \sum_{k=1}^{N} \mathbf{B}_k \vec{\mu}$;
2. *the covariance matrix of* $\mathbf{X}^*(\mathbf{u}_0)$ *is* $\mathrm{Cov}[\mathbf{X}^*(\mathbf{u}_0)] = \sum_{k=1}^{N}\sum_{\ell=1}^{N} \mathbf{B}_k \mathbf{\Sigma}(\mathbf{u}_\ell - \mathbf{u}_k)\mathbf{B}'_\ell$;
3. *the vector of estimation errors is* $\mathrm{E}[\mathbf{X}^*(\mathbf{u}_0) - \mathbf{X}(\mathbf{u}_0)] = \mathbf{c} + \left(\sum_{k=1}^{N} \mathbf{B}_k - \mathbf{I}\right)\vec{\mu}$;
4. *the covariance matrix of estimation errors is*

$$\begin{aligned}\mathrm{Cov}[\mathbf{X}^*(\mathbf{u}_0) - \mathbf{X}(\mathbf{u}_0)] &= \sum_{k=1}^{N}\sum_{\ell=1}^{N} \mathbf{B}_k \mathbf{\Sigma}(\mathbf{u}_\ell - \mathbf{u}_k)\mathbf{B}'_\ell + \mathbf{\Sigma}(0) \\ &\quad - \sum_{k=1}^{N} (\mathbf{B}_k \mathbf{\Sigma}(\mathbf{u}_0 - \mathbf{u}_k) + \mathbf{\Sigma}'(\mathbf{u}_0 - \mathbf{u}_k)\mathbf{B}'_k);\end{aligned}$$

5. $\mathbf{X}^*(\mathbf{u}_0)$ *is unbiased if and only if* $\mathbf{c} = (\mathbf{I} - \sum_{k=1}^{N} \mathbf{B}_k)\vec{\mu}$.

Proof. The proof of item 1 follows from the linearity of the expectation and of the estimator defined in equation (5.1), as well as from the second-order stationarity of the vector random function.

The proof of item 2 is derived using equation (5.2) and the fact that $\mathrm{Cov}(\mathbf{c} + \mathbf{B}\mathbf{X}) = \mathbf{B}\mathrm{Cov}(\mathbf{X})\mathbf{B}'$ for any vector of constants $\mathbf{c}$, any matrix of coefficients $\mathbf{B}$, and any random vector $\mathbf{X}$. Proof of item 3 is obtained from the first property and from $\mathrm{E}[\mathbf{X}(\mathbf{u}_0)] = \vec{\mu}$. The proof of item 4 is derived using equation (5.2), writing

$$\begin{aligned}\mathrm{Cov}[\mathbf{X}^*(\mathbf{u}_0) - \mathbf{X}(\mathbf{u}_0)] &= \mathrm{Cov}[\mathbf{B}\mathbf{X}] + \mathrm{Cov}[\mathbf{X}(\mathbf{u}_0)] \\ &\quad - \mathrm{Cov}[\mathbf{B}\mathbf{X}, \mathbf{X}(\mathbf{u}_0)]\,(\mathrm{Cov}[\mathbf{B}\mathbf{X}, \mathbf{X}(\mathbf{u}_0)])',\end{aligned}$$

and considering the fact that $\mathrm{Cov}[\mathbf{B}\mathbf{X}, \mathbf{X}(\mathbf{u}_0)] = \sum_{k=1}^{N} \mathbf{B}_k \mathbf{\Sigma}(\mathbf{u}_0 - \mathbf{u}_k)$. □

If symmetry of $\mathbf{\Sigma}(\mathbf{h})$ is assumed, no essential differences in the equations arise because $\mathbf{\Sigma}(\mathbf{h}) = \mathbf{\Sigma}'(\mathbf{h})$ does not imply $\mathbf{B}_k = \mathbf{B}'_k$, nor does it imply

$$\mathbf{B}_k \mathbf{\Sigma}(\mathbf{u}_0 - \mathbf{u}_k) = \mathbf{\Sigma}'(\mathbf{u}_0 - \mathbf{u}_k)\mathbf{B}'_k\,, \tag{5.3}$$

or

$$\text{Cov}[\mathbf{BX}, \mathbf{X}(\mathbf{u}_0)] = \text{Cov}[\mathbf{X}(\mathbf{u}_0), \mathbf{BX}]. \tag{5.4}$$

Nevertheless,

$$\mathbf{B}_k \mathbf{\Sigma}(\mathbf{u}_0 - \mathbf{u}_k) = \mathbf{\Sigma}'(\mathbf{u}_0 - \mathbf{u}_k)\mathbf{B}'_k \Longrightarrow \text{Cov}[\mathbf{BX}, \mathbf{X}(\mathbf{u}_0)] = \text{Cov}[\mathbf{X}(\mathbf{u}_0), \mathbf{BX}],$$

while the converse is not necessarily true unless it holds for any sample size N; and then, for $N = 1$, it is trivial. If either of the assumptions (5.3) or (5.4) is true, then item 4 of property 5.1 may be rewritten as

$$\text{Cov}[\mathbf{X}^*(\mathbf{u}_0) - \mathbf{X}(\mathbf{u}_0)] = \sum_{k=1}^{N}\sum_{\ell=1}^{N} \mathbf{B}_k \mathbf{\Sigma}(\mathbf{u}_\ell - \mathbf{u}_k)\mathbf{B}'_\ell + \mathbf{\Sigma}(0) - 2\sum_{k=1}^{N} \mathbf{B}_k \mathbf{\Sigma}(\mathbf{u}_0 - \mathbf{u}_k).$$

These assumptions do not change the cokriging estimator, however, nor the cokriging equations, nor the estimation variance discussed hereafter.

As can be seen from item 4 of property 5.1, a covariance matrix of estimation errors exists. The concept of *estimation variance* for $\mathbf{X}(\mathbf{u}_0)$, that is, the value to be minimized when looking for an estimator with minimum estimation variance, was defined by Myers (1982) as the trace of the matrix mentioned above:

$$\sigma_e^2(\mathbf{u}_0) = \text{Tr}\left(\text{Cov}[\mathbf{X}^*(\mathbf{u}_0) - \mathbf{X}(\mathbf{u}_0)]\right). \tag{5.5}$$

From equation 5.5, item 4 of property 5.1, and the fact that for any matrix $\mathbf{M}$, $\text{Tr}(\mathbf{M}) = \text{Tr}(\mathbf{M}')$, the following equation results

$$\begin{aligned}\sigma_e^2(\mathbf{u}_0) \quad &= \quad \sum_{k=1}^{N}\sum_{\ell=1}^{N} \text{Tr}\left(\mathbf{B}_k \mathbf{\Sigma}(\mathbf{u}_\ell - \mathbf{u}_k)\mathbf{B}'_\ell\right) + \text{Tr}\left(\mathbf{\Sigma}(0)\right) \\ &\quad -2\sum_{k=1}^{N} \text{Tr}\left(\mathbf{B}_k \mathbf{\Sigma}(\mathbf{u}_0 - \mathbf{u}_k)\right).\end{aligned}$$

This equation is the most suitable form for defining an estimation variance in the usual sense, not only because it is simpler to treat computationally, but also because the estimation variances for the individual components can be derived.

5.1.1 Cokriging with known mean

If the vector of expected values $\vec{\mu}$ is known, then the vector of constants in equation 5.2 is

$$\mathbf{c} = (\mathbf{I} - \sum_{k=1}^{N} \mathbf{B}_k)\vec{\mu}\,.$$

The resulting estimator,

$$\mathbf{X}^*(\mathbf{u}_0) = \vec{\mu} + \sum_{k=1}^{N} \mathbf{B}_k(\mathbf{X}(\mathbf{u}_k) - \vec{\mu}),$$

is unbiased for any set of matrices $\{\mathbf{B}_k; k = 1, 2, \ldots, N\}$. Consequently, the matrices of weights are not subject to constraints when unbiasedness is stipulated. From the resulting set of linear unbiased estimators for $\mathbf{X}(\mathbf{u}_0)$, the one with the smallest estimation variance should be chosen. Results stated in properties 5.2 and 5.3 are due to Myers (1982).

Property 5.2 *If the vector of expected values $\vec{\mu}$ is known, then the estimation variance, $\sigma_e^2(\mathbf{u}_0)$, reaches a minimum when the set of matrices $\{\mathbf{B}_k;$ $k = 1, 2, \ldots, N\}$ satisfies the system of equations:*

$$\sum_{k=1}^{N} \mathbf{B}_k \boldsymbol{\Sigma}(\mathbf{u}_\ell - \mathbf{u}_k) - \boldsymbol{\Sigma}(\mathbf{u}_\ell - \mathbf{u}_0) = 0, \quad \forall\, \ell = 1, 2, \ldots, N.$$

Proof. In order to determine an extreme value, the first derivatives of the function under consideration are set equal to zero. The notation used for symbolic vector and matrix differentiation and the derivative rules are taken from Fahrmeir and Hamerle (1984). From equation (5.5) it holds that

$$\frac{\partial \sigma_e^2(\mathbf{u}_0)}{\partial \mathbf{B}_\ell} = \sum_{k=1}^{N}\sum_{j=1}^{N} \frac{\partial \mathrm{Tr}\,(\mathbf{B}_k \boldsymbol{\Sigma}(\mathbf{u}_j - \mathbf{u}_k)\mathbf{B}_j')}{\partial \mathbf{B}_\ell} \tag{5.6}$$

$$-2\sum_{k=1}^{N} \frac{\partial \mathrm{Tr}\,(\mathbf{B}_k \boldsymbol{\Sigma}(\mathbf{u}_0 - \mathbf{u}_k))}{\partial \mathbf{B}_\ell} \tag{5.7}$$

$$+\frac{\partial \mathrm{Tr}\,(\boldsymbol{\Sigma}(0))}{\partial \mathbf{B}_\ell}. \tag{5.8}$$

Considering each one of the three terms in the sum separately, and the fact that $\boldsymbol{\Sigma}(\mathbf{h}) = \boldsymbol{\Sigma}'(-\mathbf{h})$ [see equation (5.1)], we obtain for part (5.6)

$$\begin{aligned}
\sum_{k=1}^{N}\sum_{j=1}^{N} & \frac{\partial \mathrm{Tr}\left(\mathbf{B}_k \boldsymbol{\Sigma}(\mathbf{u}_j - \mathbf{u}_k)\mathbf{B}_j'\right)}{\partial \mathbf{B}_\ell} \\
&= \sum_{j \neq \ell} \left(\boldsymbol{\Sigma}(\mathbf{u}_j - \mathbf{u}_\ell)\mathbf{B}_j'\right)' + \sum_{k \neq \ell} \mathbf{B}_k \boldsymbol{\Sigma}(\mathbf{u}_\ell - \mathbf{u}_k) \\
&\quad + \mathbf{B}_\ell \left(\boldsymbol{\Sigma}'(\mathbf{u}_\ell - \mathbf{u}_\ell) + \boldsymbol{\Sigma}(\mathbf{u}_\ell - \mathbf{u}_\ell)\right) \\
&= \sum_{k=1}^{N} \mathbf{B}_k \left(\boldsymbol{\Sigma}'(\mathbf{u}_k - \mathbf{u}_\ell) + \boldsymbol{\Sigma}(\mathbf{u}_\ell - \mathbf{u}_k)\right) \\
&= 2\sum_{k=1}^{N} \mathbf{B}_k \boldsymbol{\Sigma}(\mathbf{u}_\ell - \mathbf{u}_k);
\end{aligned}$$

for part (5.7) the result is

$$\begin{aligned}
-2\sum_{k=1}^{N} & \frac{\partial \mathrm{Tr}\left(\mathbf{B}_k \boldsymbol{\Sigma}(\mathbf{u}_0 - \mathbf{u}_k)\right)}{\partial \mathbf{B}_\ell} \\
&= -2\boldsymbol{\Sigma}'(\mathbf{u}_0 - \mathbf{u}_\ell) \\
&= -2\boldsymbol{\Sigma}(\mathbf{u}_\ell - \mathbf{u}_0);
\end{aligned}$$

while for part (5.8) the result is

$$\frac{\partial \mathrm{Tr}\left(\boldsymbol{\Sigma}(0)\right)}{\partial \mathbf{B}_\ell} = 0.$$

Therefore,

$$\frac{\partial \sigma_e^2(\mathbf{u}_0)}{\partial \mathbf{B}_\ell} = 2\sum_{k=1}^{N} \mathbf{B}_k \boldsymbol{\Sigma}(\mathbf{u}_\ell - \mathbf{u}_k) - 2\boldsymbol{\Sigma}(\mathbf{u}_\ell - \mathbf{u}_0)$$

for all $\ell = 1, 2, \ldots, N$. Equating these results to zero and simplifying, the desired system of equations is obtained. □

The matrices of weights determined in this manner lead to a linear unbiased estimator $\mathbf{X}^*(\mathbf{u}_0)$ of $\mathbf{X}(\mathbf{u}_0)$ which minimizes the estimation variance, $\sigma_e^2(\mathbf{u}_0)$. This estimator is called—in analogy to the terminology used for kriging—the *cokriging estimator with known mean*, or the *simple cokriging estimator*, and is written $\mathbf{X}_s^*(\mathbf{u}_0)$.

Property 5.3 *The estimation variance of the cokriging estimator with known mean is*

$$\sigma_e^2(\mathbf{u}_0) = \mathrm{Tr}\left(\boldsymbol{\Sigma}(0)\right) - \sum_{k=1}^{N} \mathrm{Tr}\left(\mathbf{B}_k \boldsymbol{\Sigma}(\mathbf{u}_0 - \mathbf{u}_k)\right).$$

Proof. If the $\mathbf{B}_k$'s satisfy the system of equations in property 5.2,

$$\sum_{k=1}^{N} \mathbf{B}_k \boldsymbol{\Sigma}(\mathbf{u}_\ell - \mathbf{u}_k) = \boldsymbol{\Sigma}(\mathbf{u}_\ell - \mathbf{u}_0) = \boldsymbol{\Sigma}'(\mathbf{u}_0 - \mathbf{u}_\ell)$$

for all $\ell = 1, 2, \ldots, N$; then, multiplying each equation by the corresponding $\mathbf{B}'_\ell$ and summing over ℓ,

$$\sum_{\ell=1}^{N} \sum_{k=1}^{N} \mathbf{B}_k \boldsymbol{\Sigma}(\mathbf{u}_\ell - \mathbf{u}_k)\mathbf{B}'_\ell = \sum_{\ell=1}^{N} \boldsymbol{\Sigma}'(\mathbf{u}_0 - \mathbf{u}_\ell)\mathbf{B}'_\ell . \tag{5.9}$$

The trace is linear and for any matrix $\mathbf{M}$, $\mathrm{Tr}(\mathbf{M}) = \mathrm{Tr}(\mathbf{M}')$, so the desired result is produced by substitution of equation (5.9) into equation (5.8) and subsequent simplification. The same result may be obtained by substitution of equation (5.9) into item 4 of property 5.1, and posterior consideration of the trace. □

The estimation variance of the simple cokriging estimator, obtained in property 5.3, is denoted by $\sigma_s^2(\mathbf{u}_0)$ and the simple cokriging estimation error is then defined to be $\sigma_s(\mathbf{u}_0)$.

To determine the estimation variances of individual components, Myers (1982) noted that:

$$\mathrm{Cov}[\mathbf{X}^*(\mathbf{u}_0) - \mathbf{X}(\mathbf{u}_0)] = \left(\mathrm{Cov}[(X_i^*(\mathbf{u}_0) - X_i(\mathbf{u}_0)), (X_j^*(\mathbf{u}_0) - X_j(\mathbf{u}_0))]\right),$$

and, writing

$$\mathrm{Cov}[(X_i^*(\mathbf{u}_0) - X_i(\mathbf{u}_0)), (X_j^*(\mathbf{u}_0) - X_j(\mathbf{u}_0))] = v_{ij}(\mathbf{u}_0),$$

it is

$$\mathrm{Cov}[\mathbf{X}^*(\mathbf{u}_0) - \mathbf{X}(\mathbf{u}_0)] = (\, v_{ij}(\mathbf{u}_0)\,) = \boldsymbol{\Sigma}_s(\mathbf{u}_0). \tag{5.10}$$

Therefore,

$$\sigma_s^2(\mathbf{u}_0) = \mathrm{Tr}(\boldsymbol{\Sigma}_s(\mathbf{u}_0)) = \sum_{i=1}^{q} v_{ii}(\mathbf{u}_0) = \sum_{i=1}^{q} \mathrm{Cov}[X_i^*(\mathbf{u}_0) - X_i(\mathbf{u}_0)]$$

and, as a consequence, it is possible to compute the matrix $\boldsymbol{\Sigma}_s(\mathbf{u}_0)$ after solution of the system of equations for minimization of $\sigma_e^2(\mathbf{u}_0)$ and, subsequently, the estimation variance of the individual components as

$$v_{ii}(\mathbf{u}_0) = \sigma_{ii}(0) - \sum_{k=1}^{N} \sum_{j=1}^{q} b_{ij}^k \sigma_{ji}(\mathbf{u}_0 - \mathbf{u}_k)$$

or as

$$v_{ii}(\mathbf{u}_0) = \sigma_{ii}(0) - \sum_{k=1}^{N}\sum_{j\neq k} b_{ij}^k \sigma_{ji}(\mathbf{u}_0 - \mathbf{u}_k) - \sum_{k=1}^{N} b_{ii}^k \sigma_{ii}(\mathbf{u}_0 - \mathbf{u}_k), \qquad (5.11)$$

with $\mathbf{B}_k = (b_{ij}^k)$. The simple kriging method for the ith variable would provide a kriging estimation variance

$$\sigma_i^2(\mathbf{u}_0) = \sigma_{ii}(0) - \sum_{k=1}^{N} \beta_k \cdot \sigma_{ii}(\mathbf{u}_0 - \mathbf{u}_k). \qquad (5.12)$$

The comparison of equation (5.12) with equation (5.10) shows that, in general, $v_{ii}(\mathbf{u}_0)$ will be smaller than $\sigma_i^2(\mathbf{u}_0)$, because the β_k's satisfy the same conditions as the b_{ii}^k's. Numerical results confirm this assertion.

5.1.2 Cokriging with unknown mean

If the vector of expected values $\vec{\mu}$ is unknown, then the vector of constants in equation (5.2) is chosen as $\mathbf{c} = 0$. This case is also discussed in Myers (1982), where the following statements are set forth.

Property 5.4 *If $\vec{\mu}$ is unknown, and $\mathbf{c} = 0$* [in equation (5.2)], *then $\mathbf{X}^*(\mathbf{u}_0)$ is unbiased if and only if $\sum_{k=1}^{N} \mathbf{B}_k = \mathbf{I}$.*

This assertion is a direct consequence of item 5 of property 5.1. Hence, the $\mathbf{B}_k$'s are subject to a constraint, and to minimize the estimation variance, Lagrange multipliers are used. At the same time, as with known mean, the estimation variance $\sigma_e^2(\mathbf{u}_0)$ is defined as the trace of the covariance matrix of the estimation errors, leading to the following statement:

Property 5.5 *If the vector of expected values $\vec{\mu}$ is unknown and the linear estimator*

$$\mathbf{X}^*(\mathbf{u}_0) = \sum_{k=1}^{N} \mathbf{B}_k \mathbf{X}(\mathbf{u}_k)$$

is unbiased, then $\sigma_e^2(\mathbf{u}_0)$ reaches a minimum if the matrices of weights $\mathbf{B}_k$, $k = 1, 2, \ldots, N$, satisfy the following system of equations:

$$\begin{aligned}
\sum_{k=1}^{N} \mathbf{B}_k \mathbf{\Sigma}(\mathbf{u}_\ell - \mathbf{u}_k) + \mathbf{\Lambda} &= \mathbf{\Sigma}(\mathbf{u}_\ell - \mathbf{u}_0), \quad \ell = 1, 2, \ldots, N; \\
\sum_{k=1}^{N} \mathbf{B}_k &= \mathbf{I},
\end{aligned}$$

where $\mathbf{\Lambda}$ is a $q \times q$ matrix of Lagrange multipliers.

Proof. The proof is analogous to the proof of property 5.2, taking into account property 5.4. Lagrange multipliers must be used to incorporate the restriction imposed on the matrices of weights by the condition of unbiasedness when minimizing equation (5.5). □

If the matrices $\mathbf{B}_k$, $k = 1, 2, \ldots, N$, satisfy the system of equations stated in property 5.5, then

$$\mathbf{X}^*(\mathbf{u}_0) = \sum_{k=1}^{N} \mathbf{B}_k \mathbf{X}(\mathbf{u}_k)$$

is the linear unbiased estimator with minimum estimation variance among the possible linear unbiased estimators which use $\mathbf{X}(\mathbf{u}_k), k = 1, 2, \ldots, N$, for the estimation of $\mathbf{X}(\mathbf{u}_0)$. It is called the *general,* or *ordinary, cokriging estimator* and is written $\mathbf{X}_o^*(\mathbf{u}_0)$. Its estimation variance may be calculated from

$$\sigma_e^2(\mathbf{u}_0) = \operatorname{Tr}(\boldsymbol{\Sigma}(0)) - \sum_{k=1}^{N} \operatorname{Tr}(\mathbf{B}_k \boldsymbol{\Sigma}(\mathbf{u}_0 - \mathbf{u}_k)) - \operatorname{Tr}(\boldsymbol{\Lambda}).$$

The resulting variance, obtained in a manner analogous to that of simple cokriging, is called the *general,* or *ordinary, cokriging estimation variance* and is indicated by $\sigma_o^2(\mathbf{u}_0)$, while $\sigma_o(\mathbf{u}_0)$ is called the *general,* or *ordinary, cokriging estimation error.*

Analogous to the case where the vector of expected values is known, the ordinary cokriging estimation variance arises as the sum of the cokriging estimation variances of the individual components, which are the diagonal elements of the matrix

$$\boldsymbol{\Sigma}_o(\mathbf{u}_0) = \boldsymbol{\Sigma}(0) - \sum_{k=1}^{N} \mathbf{B}_k \boldsymbol{\Sigma}(\mathbf{u}_0 - \mathbf{u}_k) - \boldsymbol{\Lambda}.$$

As a consequence, the individual components can be calculated after solution of the corresponding system of equations.

5.2 Normal cokriging

We now consider the case where a multivariate normal distribution can be assumed for the vector random function $\mathbf{X}(\mathbf{u})$. This means that the random vector with $q(N+1)$ elements

$$\left(\mathbf{X}'(\mathbf{u}_0) \quad \mathbf{X}'(\mathbf{u}_1) \quad \cdots \quad \mathbf{X}'(\mathbf{u}_N) \right)'$$

follows a $q(N+1)$-multivariate normal distribution, for any choice of points $\mathbf{u}_k \in \Omega$, $k = 0, 1, 2, \ldots, N$, and any sample size N.

Although the hypothesis of multivariate normality is very strong and cannot be checked in practice, evidence of strong departure from univariate normality—like skewed experimental histograms—should be understood as a sign that a transformation of the variables under study may be advisable.

Recall that in the case of multivariate normality of the vector random function, second-order stationarity and strict stationarity are equivalent.

Property 5.6 *If the vector random function* $\mathbf{X}(\mathbf{u})$ *is stationary and follows a multivariate normal distribution and the vector of expected values* $\vec{\mu}$ *is known, then*

$$\mathbf{X}_s^*(\mathbf{u}_0) = \mathrm{E}[\mathbf{X}(\mathbf{u}_0) \mid \mathbf{X}(\mathbf{u}_1), \mathbf{X}(\mathbf{u}_2), \ldots, \mathbf{X}(\mathbf{u}_N)],$$

and

$$\mathbf{\Sigma}_s(\mathbf{u}_0) = \mathrm{Cov}[\mathbf{X}(\mathbf{u}_0) \mid \mathbf{X}(\mathbf{u}_1), \mathbf{X}(\mathbf{u}_2), \ldots, \mathbf{X}(\mathbf{u}_N)].$$

That is, the simple cokriging estimator agrees with the conditional expectation of the random vector $\mathbf{X}(\mathbf{u}_0)$*, given the sample, and the covariance matrix of the estimation errors, given the sample, agrees with the conditional covariance matrix of* $\mathbf{X}(\mathbf{u}_0)$.

Proof. By definition of conditional expectation and conditional covariance of a multivariate normal random vector [see Fahrmeir and Hamerle (1984) for a detailed account],

$$\mathrm{E}[\mathbf{X}(\mathbf{u}_0) \mid \mathbf{X}(\mathbf{u}_1), \ldots, \mathbf{X}(\mathbf{u}_N)] = \vec{\mu} + \mathbf{\Sigma}_{0,1\ldots N}\mathbf{\Sigma}_{1\ldots N}^{-1}\left(\mathbf{X}'(\mathbf{u}_1) - \vec{\mu}' \cdots \mathbf{X}'(\mathbf{u}_N) - \vec{\mu}'\right)'$$

and

$$\mathrm{Cov}[\mathbf{X}(\mathbf{u}_0) \mid \mathbf{X}(\mathbf{u}_1), \ldots, \mathbf{X}(\mathbf{u}_N)] = \mathrm{Cov}[\mathbf{X}(\mathbf{u}_0)] - \mathbf{\Sigma}_{0,1\ldots N}\mathbf{\Sigma}_{1\ldots N}^{-1}\mathbf{\Sigma}_{1\ldots N,0},$$

where

$$\begin{aligned}
\mathbf{\Sigma}_{0,1\ldots N} &= \mathrm{Cov}\left[\mathbf{X}(\mathbf{u}_0), \left(\mathbf{X}'(\mathbf{u}_1) \quad \cdots \quad \mathbf{X}'(\mathbf{u}_N)\right)'\right], \\
\mathbf{\Sigma}_{1\ldots N,0} &= \mathrm{Cov}\left[\left(\mathbf{X}'(\mathbf{u}_1) \quad \cdots \quad \mathbf{X}'(\mathbf{u}_N)\right)', \mathbf{X}(\mathbf{u}_0)\right], \\
\mathbf{\Sigma}_{1\ldots N} &= \mathrm{Cov}\left[\left(\mathbf{X}'(\mathbf{u}_1) \quad \cdots \quad \mathbf{X}'(\mathbf{u}_N)\right)'\right].
\end{aligned}$$

Taking $\left(\mathbf{B}_1 \quad \mathbf{B}_2 \quad \cdots \quad \mathbf{B}_N\right) = \mathbf{\Sigma}_{0,1\ldots N}\mathbf{\Sigma}_{1\ldots N}^{-1}$, it holds that

$$\mathrm{E}[\mathbf{X}(\mathbf{u}_0) \mid \mathbf{X}(\mathbf{u}_1), \ldots, \mathbf{X}(\mathbf{u}_N)] = \vec{\mu} + \sum_{k=1}^{N} \mathbf{B}_k \left(\mathbf{X}(\mathbf{u}_k) - \vec{\mu}\right),$$

and the $\mathbf{B}_k$'s must satisfy the system of equations:

$$\left(\mathbf{B}_1 \quad \mathbf{B}_2 \quad \cdots \quad \mathbf{B}_N\right)\mathbf{\Sigma}_{1\ldots N} = \mathbf{\Sigma}_{0,1\ldots N}.$$

That is,

$$\begin{pmatrix} \mathbf{B}_1 & \mathbf{B}_2 & \cdots & \mathbf{B}_N \end{pmatrix} \begin{pmatrix} \mathbf{\Sigma}(\mathbf{u}_1-\mathbf{u}_1) & \mathbf{\Sigma}(\mathbf{u}_2-\mathbf{u}_1) & \cdots & \mathbf{\Sigma}(\mathbf{u}_N-\mathbf{u}_1) \\ \mathbf{\Sigma}(\mathbf{u}_1-\mathbf{u}_2) & \mathbf{\Sigma}(\mathbf{u}_2-\mathbf{u}_2) & \cdots & \mathbf{\Sigma}(\mathbf{u}_N-\mathbf{u}_2) \\ \vdots & \vdots & \ddots & \vdots \\ \mathbf{\Sigma}(\mathbf{u}_1-\mathbf{u}_N) & \mathbf{\Sigma}(\mathbf{u}_2-\mathbf{u}_N) & \cdots & \mathbf{\Sigma}(\mathbf{u}_N-\mathbf{u}_N) \end{pmatrix}$$

$$= \begin{pmatrix} \mathbf{\Sigma}(\mathbf{u}_1-\mathbf{u}_0) & \mathbf{\Sigma}(\mathbf{u}_2-\mathbf{u}_0) & \cdots & \mathbf{\Sigma}(\mathbf{u}_N-\mathbf{u}_0) \end{pmatrix},$$

or

$$\sum_{k=1}^{N} \mathbf{B}_k \mathbf{\Sigma}(\mathbf{u}_\ell - \mathbf{u}_k) = \mathbf{\Sigma}(\mathbf{u}_\ell - \mathbf{u}_0), \qquad \text{for all} \quad \ell = 1, 2, \ldots, N.$$

For the conditional covariance matrix it then holds that

$$\mathrm{Cov}[\mathbf{X}(\mathbf{u}_0)] = \mathbf{\Sigma}(0),$$

$$\mathbf{\Sigma}_{0,1\ldots N}\mathbf{\Sigma}_{1\ldots N}^{-1} = \begin{pmatrix} \mathbf{B}_1 & \mathbf{B}_2 & \cdots & \mathbf{B}_N \end{pmatrix},$$

$$\begin{aligned} \begin{pmatrix} \mathbf{B}_1 & \mathbf{B}_2 & \cdots & \mathbf{B}_N \end{pmatrix} \mathbf{\Sigma}_{1\ldots N,0} &= \begin{pmatrix} \mathbf{B}_1 & \mathbf{B}_2 & \cdots & \mathbf{B}_N \end{pmatrix} \begin{pmatrix} \mathbf{\Sigma}(\mathbf{u}_0-\mathbf{u}_1) \\ \mathbf{\Sigma}(\mathbf{u}_0-\mathbf{u}_2) \\ \vdots \\ \mathbf{\Sigma}(\mathbf{u}_0-\mathbf{u}_N) \end{pmatrix} \\ &= \sum_{k=1}^{N} \mathbf{B}_k \mathbf{\Sigma}(\mathbf{u}_0 - \mathbf{u}_k), \end{aligned}$$

and therefore,

$$\mathrm{Cov}[\mathbf{X}(\mathbf{u}_0) \mid \mathbf{X}(\mathbf{u}_1), \ldots, \mathbf{X}(\mathbf{u}_N)] = \mathbf{\Sigma}(0) - \sum_{k=1}^{N} \mathbf{B}_k \mathbf{\Sigma}(\mathbf{u}_0 - \mathbf{u}_k).$$

□

In view of these results, as with estimation of a single random function using kriging, it is clear that the transformation of variables (for example, taking the logarithm of variables with an experimental skewed distribution) may lead in some cases to better estimations. Therefore, it makes sense to study other multivariate distributional assumptions.

5.3 Lognormal cokriging

Assume that $\mathbf{X}(\mathbf{u})$ is a vector random function, with sample space the positive orthant of q-dimensional real space, that follows a multivariate lognormal distribution. The multivariate lognormal distribution is supposed to depend on one vector- and one matrix-valued parameter related to the vector of expected values and to the matrix of auto- and cross-covariance functions. The case of the three-parameter lognormal distribution—which in this situation would involve another vector-valued parameter—is not considered explicitly, since it can be transformed to the two-parameter case by a simple translation. Nevertheless, it should be pointed out that estimation of the additional vector-valued parameter is not an easy task, even in the single-variable case; refer to Rendu (1979) for further details. Second-order log-stationarity and the existence of a sample or partial realization $\mathbf{X}(\mathbf{u}_k)$, $k = 1, 2, \ldots, N$, are assumed.

For subsequent developments, some essential equations of the multivariate lognormal distribution are required. The bivariate lognormal distribution can be found in Aitchison and Brown (1957), from which the following properties can be derived in a straightforward manner.

Property 5.7 *If $\mathbf{X}(\mathbf{u})$ is a second-order log-stationary, multivariate lognormally distributed vector random function with*

$$\mathrm{E}[\ln(\mathbf{X}(\mathbf{u}))] = \vec{\mu} = (\mu_1, \mu_2, \ldots, \mu_q)'$$

and

$$\mathrm{Cov}[\ln(\mathbf{X}(\mathbf{u})), \ln(\mathbf{X}(\mathbf{u}+\mathbf{h}))] = \mathbf{\Sigma}(\mathbf{h}) = (\,\sigma_{ij}(\mathbf{h})\,)\,,$$

for any $\mathbf{u} \in \Omega$ and every lag $\mathbf{h}$, then

1. $\mathrm{E}[\mathbf{X}(\mathbf{u})] = \exp\left(\vec{\mu} + \frac{1}{2}\mathrm{diag}\,(\mathbf{\Sigma}(0))\,\mathbf{j}\right)$,
2. $\mathrm{Cov}[\mathbf{X}(\mathbf{u}), \mathbf{X}(\mathbf{u}+\mathbf{h})] = \mathrm{diag}\,(\mathrm{E}[\mathbf{X}(\mathbf{u})])\,(\exp(\mathbf{\Sigma}(\mathbf{h})) - \mathbf{j}\mathbf{j}')\,\mathrm{diag}\,(\mathrm{E}[\mathbf{X}(\mathbf{u})])$.

Proof. Since for each $\mathbf{u} \in \Omega$ every component $X_i(\mathbf{u})$ of $\mathbf{X}(\mathbf{u})$ is lognormally distributed, under the hypothesis of second-order log-stationarity it holds that $\mathrm{E}[X_i(\mathbf{u})] = \exp(\mu_i + \sigma_{ii}(0))$, and in matrix notation, item 1 of property 5.7 is obtained.

To prove item 2 of property 5.7, recall that

$$\mathrm{Cov}[X_i(\mathbf{u}), X_j(\mathbf{u}+\mathbf{h})] = \mathrm{E}[X_i(\mathbf{u})X_j(\mathbf{u}+\mathbf{h})] - \mathrm{E}[X_i(\mathbf{u})]\mathrm{E}[X_j(\mathbf{u}+\mathbf{h})];$$

since item 1 of property 5.7 implies that $\mathrm{E}[\mathbf{X}(\mathbf{u})]$ does not depend on $\mathbf{u}$, we can write $\mathrm{E}[\mathbf{X}(\mathbf{u})] = \vec{\alpha} = (\alpha_1, \alpha_2, \ldots, \alpha_d)'$, and thus, $\mathrm{E}[X_i(\mathbf{u})]\mathrm{E}[X_j(\mathbf{u}+\mathbf{h})] = \alpha_i\alpha_j$.

Furthermore, from the properties of the bivariate lognormal distribution, as given by Aitchison and Brown (1957),

$$\begin{aligned}
\mathrm{E}[X_i(\mathbf{u})X_j(\mathbf{u}+\mathbf{h})] &= \exp\left(\mu_i + \mu_j + \tfrac{1}{2}(\sigma_{ii}(0) + \sigma_{jj}(0) + 2\sigma_{ij}(\mathbf{h}))\right) \\
&= \exp\left(\ln(\alpha_i) - \tfrac{1}{2}\sigma_{ii}(0) + \ln(\alpha_j) - \tfrac{1}{2}\sigma_{jj}(0)\right. \\
&\quad \left. + \tfrac{1}{2}\sigma_{ii}(0) + \tfrac{1}{2}\sigma_{jj}(0) + \sigma_{ij}(\mathbf{h})\right) \\
&= \exp(\ln(\alpha_i\alpha_j) + \sigma_{ij}(\mathbf{h})) \\
&= \alpha_i\alpha_j \exp(\sigma_{ij}(\mathbf{h})).
\end{aligned}$$

Finally,

$$\begin{aligned}
\mathrm{Cov}[X_i(\mathbf{u}), X_j(\mathbf{u}+\mathbf{h})] &= \alpha_i\alpha_j \exp\left(\sigma_{ij}(\mathbf{h})\right) - \alpha_i\alpha_j \\
&= \alpha_i \left(\exp\left(\sigma_{ij}(\mathbf{h})\right) - 1\right) \alpha_j,
\end{aligned}$$

and written in matrix form, we get item 2 of property 5.7. □

The next step is to estimate the unknown $\mathbf{X}(\mathbf{u}_0)$ with a log-linear estimator:

$$\mathbf{X}^*(\mathbf{u}_0) = \exp\left(\mathbf{c} + \sum_{k=1}^{N} \mathbf{B}_k \ln(\mathbf{X}(\mathbf{u}_k))\right). \tag{5.13}$$

Property 5.8 *Under the same hypothesis as that of property 5.7, and denoting* $\mathrm{E}[\mathbf{X}(\mathbf{u})] = \vec{\alpha}$, *the following properties hold for* $\ln(\mathbf{X}^*(\mathbf{u}_0))$, *as defined in equation (5.13):*

1. $\mathrm{E}\left[\ln(\mathbf{X}^*(\mathbf{u}_0))\right] = \mathbf{c} + \sum_{k-1}^{N} \mathbf{B}_k \left(\ln(\vec{\alpha}) - \tfrac{1}{2}\mathrm{diag}(\boldsymbol{\Sigma}(0))\mathbf{j}\right)$;

2. $\mathrm{Cov}\left[\ln(\mathbf{X}^*(\mathbf{u}_0))\right] = \sum_{k=1}^{N}\sum_{\ell=1}^{N} \mathbf{B}_k \boldsymbol{\Sigma}(\mathbf{u}_\ell - \mathbf{u}_k)\mathbf{B}'_\ell$;

3. $\mathrm{E}[\mathbf{X}^*(\mathbf{u}_0)] = \exp(M)$, *where*

$$\begin{aligned}
M = \mathbf{c} &+ \sum_{k=1}^{N} \mathbf{B}_k \left(\ln(\vec{\alpha}) - \tfrac{1}{2}\mathrm{diag}(\boldsymbol{\Sigma}(0))\mathbf{j}\right) \\
&+ \tfrac{1}{2}\sum_{k=1}^{N}\sum_{\ell=1}^{N} \mathrm{diag}\left(\mathbf{B}_k \boldsymbol{\Sigma}(\mathbf{u}_\ell - \mathbf{u}_k)\mathbf{B}'_\ell\right)\mathbf{j};
\end{aligned}$$

4. $\mathrm{Cov}[\mathbf{X}^*(\mathbf{u}_0)] = \mathrm{diag}\left(\mathrm{E}[\mathbf{X}^*(\mathbf{u}_0)]\right)\left(\exp(M) - \mathbf{j}\mathbf{j}'\right)\mathrm{diag}\left(\mathrm{E}[\mathbf{X}^*(\mathbf{u}_0)]\right)$, *where*

$$M = \sum_{k=1}^{N}\sum_{\ell=1}^{N} \mathbf{B}_k \boldsymbol{\Sigma}(\mathbf{u}_\ell - \mathbf{u}_k)\mathbf{B}'_\ell;$$

5. $\mathbf{X}^*(\mathbf{u}_0)$ *is unbiased if and only if*

$$\begin{aligned}\mathbf{c} \;=\;& \left(\mathbf{I} - \sum_{k=1}^{N} \mathbf{B}_k\right) \ln(\vec{\alpha}) + \tfrac{1}{2} \sum_{k=1}^{N} \mathbf{B}_k \mathrm{diag}\left(\mathbf{\Sigma}(0)\right) \mathbf{j} \\ & -\tfrac{1}{2} \sum_{k=1}^{N} \sum_{\ell=1}^{N} \mathrm{diag}\left(\mathbf{B}_k \mathbf{\Sigma}(\mathbf{u}_\ell - \mathbf{u}_k) \mathbf{B}'_\ell\right) \mathbf{j};\end{aligned}$$

6. *If* $\mathbf{X}^*(\mathbf{u}_0)$ *is unbiased and* $\mathbf{X}^*(\mathbf{u}_0)$ *and* $\mathbf{X}(\mathbf{u}_0)$ *are jointly lognormally distributed, then the covariance matrix of the estimation errors is*

$$\mathrm{Cov}[\mathbf{X}^*(\mathbf{u}_0) - \mathbf{X}(\mathbf{u}_0)] = \mathrm{diag}(\vec{\alpha})(M_1 - M_2)\mathrm{diag}(\vec{\alpha}),$$

where

$$M_1 = \exp\left(\mathbf{\Sigma}(0)\right) + \exp\left(\sum_{k=1}^{N} \sum_{\ell=1}^{N} \mathbf{B}_k \mathbf{\Sigma}(\mathbf{u}_\ell - \mathbf{u}_k) \mathbf{B}'_\ell\right)$$

and

$$M_2 = \exp\left(\sum_{k=1}^{N} \mathbf{B}_k \mathbf{\Sigma}(\mathbf{u}_0 - \mathbf{u}_k)\right) + \exp\left(\sum_{k=1}^{N} \mathbf{\Sigma}'(\mathbf{u}_0 - \mathbf{u}_k) \mathbf{B}'_k\right).$$

Proof. For items 1 to 5 of property 5.8, the proof is based on the characteristics of the multivariate lognormal distribution and the second-order log-stationarity of $\mathbf{X}(\mathbf{u})$ and is straightforward. For item 6 of property 5.8, it holds that

$$\begin{aligned}&\mathrm{Cov}[\mathbf{X}^*(\mathbf{u}_0) - \mathbf{X}(\mathbf{u}_0)] \\ &\quad = \mathrm{Cov}[\mathbf{X}^*(\mathbf{u}_0)] + \mathrm{Cov}[\mathbf{X}(\mathbf{u}_0)] \\ &\quad\quad - \mathrm{Cov}[\mathbf{X}^*(\mathbf{u}_0), \mathbf{X}(\mathbf{u}_0)] - \mathrm{Cov}[\mathbf{X}(\mathbf{u}_0), \mathbf{X}^*(\mathbf{u}_0)].\end{aligned} \tag{5.14}$$

Item 4 of property 5.8 gives $\mathrm{Cov}[\mathbf{X}^*(\mathbf{u}_0)]$, in terms of $\mathrm{E}[\mathbf{X}^*(\mathbf{u}_0)]$, which must be substituted by $\vec{\alpha}$ under the hypothesis of unbiasedness. Item 2 of property 5.7 gives $\mathrm{Cov}[\mathbf{X}(\mathbf{u}_0)]$, taking $\mathbf{h} = 0$. Finally, under the assumption of joint multivariate lognormality of $\mathbf{X}^*(\mathbf{u}_0)$ and $\mathbf{X}(\mathbf{u}_0)$ and unbiasedness of $\mathbf{X}^*(\mathbf{u}_0)$:

$$\begin{aligned}\mathrm{Cov}[\mathbf{X}^*(\mathbf{u}_0), \mathbf{X}(\mathbf{u}_0)] &= \mathrm{diag}(\vec{\alpha})\left(\exp\left(\mathrm{Cov}[\ln(\mathbf{X}^*(\mathbf{u}_0)), \ln(\mathbf{X}(\mathbf{u}_0))]\right) - \mathbf{j}\mathbf{j}'\right) \mathrm{diag}(\vec{\alpha}) \\ &= \mathrm{diag}(\vec{\alpha}) \left(\exp\left(\sum_{k=1}^{N} \mathbf{B}_k \mathbf{\Sigma}(\mathbf{u}_k - \mathbf{u}_0)\right) - \mathbf{j}\mathbf{j}'\right) \mathrm{diag}(\vec{\alpha}).\end{aligned}$$

The desired result is produced by substitution into equation (5.14). □

5.3.1 Lognormal cokriging with known mean

Assume the vector of expected values $\mathrm{E}[\mathbf{X}(\mathbf{u})] = \vec{\alpha}$ is known. The estimator

$$\mathbf{X}^*(\mathbf{u}_0) = \exp\left(\mathbf{c} + \sum_{k=1}^{N} \mathbf{B}_k \ln(\mathbf{X}(\mathbf{u}_k))\right), \tag{5.15}$$

with $\mathbf{c}$ defined as in item 5 of property 5.8, is unbiased for any set of weighting matrices $\{\mathbf{B}_k; k = 1, 2, \ldots, N\}$. From simple cokriging, it is known that the estimation variance of the logarithms

$$\sigma_e^2 = \mathrm{Tr}\big(\mathrm{Cov}\left[\ln(\mathbf{X}^*(\mathbf{u}_0)) - \ln(\mathbf{X}(\mathbf{u}_0))\right]\big) \tag{5.16}$$

is minimized if the $\mathbf{B}_k$'s satisfy the system of equations

$$\sum_{k=1}^{N} \mathbf{B}_k \boldsymbol{\Sigma}(\mathbf{u}_\ell - \mathbf{u}_k) = \boldsymbol{\Sigma}(\mathbf{u}_\ell - \mathbf{u}_0), \qquad \ell = 1, 2, \ldots, N. \tag{5.17}$$

As in the univariate case, the unbiased estimator obtained by substituting the $\mathbf{B}_k$'s resulting from the solution of equation (5.17) into equation (5.15) is referred to as the *lognormal cokriging estimator with known mean* or the *simple lognormal cokriging estimator*, and is denoted by $\mathbf{X}^*_{s\ell}(\mathbf{u}_0)$.

Recall that if the $\mathbf{B}_k$'s satisfy equation (5.17), then

$$\sum_{\ell=1}^{N}\sum_{k=1}^{N} \mathbf{B}_k \boldsymbol{\Sigma}(\mathbf{u}_\ell - \mathbf{u}_k)\mathbf{B}'_\ell = \sum_{\ell=1}^{N} \boldsymbol{\Sigma}(\mathbf{u}_\ell \mathbf{u}_0)\mathbf{B}'_\ell,$$

$$\boldsymbol{\Sigma}_e^2(\mathbf{u}_0) = \boldsymbol{\Sigma}(0) - \sum_{k=1}^{N} \mathbf{B}_k \boldsymbol{\Sigma}(\mathbf{u}_0 - \mathbf{u}_k),$$

and

$$\sigma_e^2(\mathbf{u}_0) = \mathrm{Tr}\left(\boldsymbol{\Sigma}(0)\right) - \sum_{k=1}^{N} \mathrm{Tr}\left(\mathbf{B}_k \boldsymbol{\Sigma}(\mathbf{u}_0 - \mathbf{u}_k)\right).$$

By simple substitution, the following properties are obtained:

Property 5.9 *The simple lognormal cokriging estimator is*

$$\mathbf{X}^*_{s\ell}(\mathbf{u}_0) = \exp\left(\left(\mathbf{I} - \sum_{k=1}^{N} \mathbf{B}_k\right)\ln(\vec{\alpha}) + \sum_{k=1}^{N} \mathbf{B}_k \ln(\mathbf{X}(\mathbf{u}_k)) + \tfrac{1}{2}\sum_{k=1}^{N} M_k\right),$$

with $M_k = \mathbf{B}_k \mathrm{diag}(\boldsymbol{\Sigma}(0))\mathbf{j} - \mathrm{diag}(\mathbf{B}_k \boldsymbol{\Sigma}(\mathbf{u}_0 - \mathbf{u}_k))\mathbf{j}$.

Property 5.10 *If* $\mathbf{X}^*(\mathbf{u}_0)$ *and* $\mathbf{X}(\mathbf{u}_0)$ *are jointly lognormally distributed, then the covariance matrix of the estimation errors is*

$$\mathrm{Cov}[\mathbf{X}^*_{s\ell}(\mathbf{u}_0) - \mathbf{X}(\mathbf{u}_0)] = \mathbf{\Sigma}_{s\ell}(\mathbf{u}_0) = \mathrm{diag}(\vec{\alpha})\mathbf{M}\mathrm{diag}(\vec{\alpha}),$$

with $\mathbf{M} = \exp(\mathbf{\Sigma}(0)) - \exp\left(\sum_{k=1}^{N} \mathbf{B}_k \mathbf{\Sigma}(\mathbf{u}_0 - \mathbf{u}_k)\right)$.

The *simple lognormal cokriging estimation variance* is $\sigma^2_{s\ell}(\mathbf{u}_0) = \mathrm{Tr}\,(\mathbf{\Sigma}_{s\ell}(\mathbf{u}_0))$ and the *simple lognormal cokriging estimation error* is $\sigma_{s\ell}(\mathbf{u}_0)$. The estimation variances of the individual components are the diagonal elements of the matrix $\mathbf{\Sigma}_{s\ell}(\mathbf{u}_0)$.

As in the case of a multivariate normally distributed vector random function, it can be shown that the conditional probability distribution, given a sample, of a multivariate lognormally distributed vector random function with known mean is lognormal. Furthermore, it holds that the conditional expectation agrees with the simple lognormal cokriging estimator $\mathbf{X}^*_{s\ell}(\mathbf{u}_0)$, and the conditional covariance matrix is

$$\begin{aligned}&\mathrm{Cov}[\mathbf{X}(\mathbf{u}_0) \mid \mathbf{X}(\mathbf{u}_1), \mathbf{X}(\mathbf{u}_2), \ldots, \mathbf{X}(\mathbf{u}_N)] \\ &= \mathrm{diag}(\vec{\alpha})\left(\exp\left(\mathbf{\Sigma}(0) - \sum_{k=1}^{N} \mathbf{B}_k \mathbf{\Sigma}(\mathbf{u}_0 - \mathbf{u}_k)\right) - \mathbf{j}\mathbf{j}'\right)\mathrm{diag}(\vec{\alpha}).\end{aligned}$$

The proof proceeds analogously to that for normal cokriging considering the relationship between expected values, covariances, and the parameters of a multivariate lognormal distribution, and it is therefore omitted.

5.3.2 Lognormal cokriging with unknown mean

If the vector of expected values is unknown, from item 5 of property 5.8,

$$\mathbf{X}^*(\mathbf{u}_0) = \exp\left(\mathbf{c} + \sum_{k=1}^{N} \mathbf{B}_k \ln\left(\mathbf{X}(\mathbf{u}_k)\right)\right) \tag{5.18}$$

is unbiased if the following conditions are satisfied:

1. $\sum_{k=1}^{N} \mathbf{B}_k = \mathbf{I}$,

2. $\mathbf{c} = \frac{1}{2}\mathrm{diag}(\mathbf{\Sigma}(0))\mathbf{j} - \frac{1}{2}\sum_{k=1}^{N}\sum_{\ell=1}^{N} \mathrm{diag}\left(\mathbf{B}_k \mathbf{\Sigma}(\mathbf{u}_\ell - \mathbf{u}_k)\mathbf{B}'_\ell\right)\mathbf{j}$

and the estimation variance to be minimized is, as in equation (5.16),

$$\sigma_e^2(\mathbf{u}_0) = \mathrm{Tr}(\boldsymbol{\Sigma}_e(\mathbf{u}_0)) = \mathrm{Tr}\big(\mathrm{Cov}\left[\ln(\mathbf{X}^*(\mathbf{u}_0)) - \ln(\mathbf{X}(\mathbf{u}_0))\right]\big).$$

In order to minimize this variance a system of equations must be solved, as for cokriging with unknown mean:

$$\sum_{k=1}^{N} \mathbf{B}_k \boldsymbol{\Sigma}(\mathbf{u}_\ell - \mathbf{u}_k) + \boldsymbol{\Lambda} = \boldsymbol{\Sigma}(\mathbf{u}_\ell - \mathbf{u}_0), \quad \forall \ell = 1, 2, \ldots, N, \tag{5.19}$$

$$\sum_{k=1}^{N} \mathbf{B}_k = \mathbf{I}, \tag{5.20}$$

where $\boldsymbol{\Lambda}$ is a $(D-1) \times (D-1)$ matrix of Lagrange multipliers. It follows that

$$\boldsymbol{\Sigma}_e(\mathbf{u}_0) = \boldsymbol{\Sigma}(0) - \sum_{k=1}^{N} \mathbf{B}_k \boldsymbol{\Sigma}(\mathbf{u}_0 - \mathbf{u}_k) - \boldsymbol{\Lambda}. \tag{5.21}$$

The unbiased estimator determined by substituting the $\mathbf{B}_k$'s obtained from the system of equations (5.20) into conditions (1) and (2) is called the *lognormal cokriging estimator with unknown mean* or the *ordinary lognormal cokriging estimator*, and is denoted by $\mathbf{X}^*_{o\ell}(\mathbf{u}_0)$.

Property 5.11 *The ordinary lognormal cokriging estimator can be written*

$$\mathbf{X}^*_{o\ell}(\mathbf{u}_0) = \exp\left(\sum_{k=1}^{N} \mathbf{B}_k \ln(\mathbf{X}(\mathbf{u}_k)) + \tfrac{1}{2}\,\mathrm{diag}\left(\boldsymbol{\Sigma}(0) - \sum_{k=1}^{N} \mathbf{B}_k \boldsymbol{\Sigma}(\mathbf{u}_0 - \mathbf{u}_k) + \boldsymbol{\Lambda}\right)\mathbf{j}\right),$$

or, alternatively,

$$\mathbf{X}^*_{o\ell}(\mathbf{u}_0) = \exp\left(\sum_{k=1}^{N} \mathbf{B}_k \ln(\mathbf{X}(\mathbf{u}_k)) + \tfrac{1}{2}\mathrm{diag}\,(\boldsymbol{\Sigma}_e(\mathbf{u}_0) + 2\boldsymbol{\Lambda})\,\mathbf{j}\right).$$

Proof. Analogously to the proof of property 5.3, using equation (5.20), the following result is obtained:

$$\sum_{k=1}^{N}\sum_{\ell=1}^{N} \mathbf{B}_k \boldsymbol{\Sigma}(\mathbf{u}_\ell - \mathbf{u}_k)\mathbf{B}'_\ell = \sum_{\ell=1}^{N} \boldsymbol{\Sigma}'(\mathbf{u}_0 - \mathbf{u}_\ell)\mathbf{B}'_\ell - \boldsymbol{\Lambda}.$$

The first equality is deduced by substituting this result into equation (5.18) and the expression thus obtained for the vector of constants $\mathbf{c}$ into (5.17). From this first equality, using (5.21), the second equality results. □

This result is of particular interest for programming aspects because it indicates that the following steps are necessary for calculation of $\mathbf{X}^*_{o\ell}(\mathbf{u}_0)$:

1. solution of the ordinary cokriging system of equations for the log–transformed variables,

2. estimation of the logarithmic values $\ln(\mathbf{X}^*(\mathbf{u}_0))$ and calculation of the associated matrix $\mathbf{\Sigma}_e(\mathbf{u}_0)$ using equation (5.21),

3. calculation of $\mathbf{X}^*_{o\ell}(\mathbf{u}_0)$ from

$$\mathbf{X}^*_{o\ell}(\mathbf{u}_0) = \exp\left(\ln(\mathbf{X}^*(\mathbf{u}_0)) + \tfrac{1}{2}\operatorname{diag}\left(\mathbf{\Sigma}_e(\mathbf{u}_0) + 2\mathbf{\Lambda}\right)\mathbf{j}\right),$$

4. calculation of the *ordinary lognormal cokriging estimation variance*

$$\sigma^2_{o\ell}(\mathbf{u}_0) = \operatorname{Tr}\left(\mathbf{\Sigma}_{o\ell}(\mathbf{u}_0)\right) = \operatorname{Tr}\left(\operatorname{Cov}[\mathbf{X}^*_{o\ell}(\mathbf{u}_0) - \mathbf{X}(\mathbf{u}_0)]\right), \qquad (5.22)$$

and/or of the *ordinary lognormal cokriging estimation error*

$$\sigma_{o\ell}(\mathbf{u}_0) = \sqrt{\operatorname{Tr}\left(\mathbf{\Sigma}_{o\ell}(\mathbf{u}_0)\right)}\,. \qquad (5.23)$$

For the computation of the variance in (5.23), recall that by item 6 of property 5.8,

$$\mathbf{\Sigma}_{o\ell}(\mathbf{u}_0) = \operatorname{diag}(\vec{\alpha})(M_1 - M_2)\operatorname{diag}(\vec{\alpha}), \qquad (5.24)$$

where

$$M_1 = \exp\left(\mathbf{\Sigma}(0)\right) + \exp\left(\sum_{k=1}^{N}\sum_{\ell=1}^{N}\mathbf{B}_k\mathbf{\Sigma}(\mathbf{u}_\ell - \mathbf{u}_k)\mathbf{B}'_\ell\right)$$

and

$$M_2 = \exp\left(\sum_{k=1}^{N}\mathbf{B}_k\mathbf{\Sigma}(\mathbf{u}_0 - \mathbf{u}_k)\right) + \exp\left(\sum_{\ell=1}^{N}\mathbf{\Sigma}'(\mathbf{u}_0 - \mathbf{u}_\ell)\mathbf{B}'_k\right).$$

If the $\mathbf{B}_k$'s satisfy the system of equations (5.20), then

$$\sum_{k=1}^{N}\sum_{\ell=1}^{N}\mathbf{B}_k\,\mathbf{\Sigma}(\mathbf{u}_\ell - \mathbf{u}_k)\,\mathbf{B}'_\ell = \sum_{k=1}^{N}\mathbf{B}_k\,\mathbf{\Sigma}(\mathbf{u}_0 - \mathbf{u}_k) - \mathbf{\Lambda}', \qquad (5.25)$$

and further, from equation (5.23),

$$\sum_{k=1}^{N}\mathbf{B}_k\,\mathbf{\Sigma}(\mathbf{u}_0 - \mathbf{u}_k) = \mathbf{\Sigma}(0) - \mathbf{\Lambda} - \mathbf{\Sigma}_e(\mathbf{u}_0)\,. \qquad (5.26)$$

Substituting (5.26) into (5.25) and then into (5.24),

$$\mathbf{\Sigma}_{o\ell}(\mathbf{u}_0) = \operatorname{diag}(\vec{\alpha})(\mathbf{M}_1 - \mathbf{M}_2)\operatorname{diag}(\vec{\alpha}), \qquad (5.27)$$

where

$$\mathbf{M}_1 = \exp(\boldsymbol{\Sigma}(0)) + \exp\left(\boldsymbol{\Sigma}(0) - \boldsymbol{\Lambda} - \boldsymbol{\Lambda}' - \boldsymbol{\Sigma}_e(\mathbf{u}_0)\right)$$

and

$$\mathbf{M}_2 = \exp\left(\boldsymbol{\Sigma}(0) - \boldsymbol{\Lambda} - \boldsymbol{\Sigma}_e(\mathbf{u}_0)\right) + \exp\left(\boldsymbol{\Sigma}(0) - \boldsymbol{\Lambda}'\boldsymbol{\Sigma}_e(\mathbf{u}_0)\right).$$

To compute the global estimation variance, $\sigma^2_{o\ell}(\mathbf{u}_0)$, the trace of the matrix $\boldsymbol{\Sigma}_{o\ell}(\mathbf{u}_0)$ is needed, and the estimation variances of each component are the summands of the trace. To obtain them, because no matrix product is involved except for the two diagonal matrices, only the diagonal elements of the sum of the matrices are of interest. Then, as only diagonal matrices will be involved, they commute; and finally, for any two matrices $\mathbf{M}_1$ and $\mathbf{M}_2$, it holds that

$$\operatorname{diag}\left(\exp(\mathbf{M}_1 + \mathbf{M}_2)\right) = \operatorname{diag}\left(\exp(\mathbf{M}_1)\right)\operatorname{diag}\left(\exp(\mathbf{M}_2)\right),$$

so the following equation results:

$$\begin{aligned}\operatorname{Tr}\left(\boldsymbol{\Sigma}_{o\ell}(\mathbf{u}_0)\right) &= \operatorname{Tr}\left(\operatorname{diag}\left(\boldsymbol{\Sigma}_{o\ell}(\mathbf{u}_0)\right)\right)\\ &= \operatorname{Tr}\left(\operatorname{diag}(\vec{\alpha})\operatorname{diag}(\vec{\alpha})\operatorname{diag}\left(\exp(\boldsymbol{\Sigma}(0))\right)(\mathbf{I} + \mathbf{M}_1\mathbf{M}_2)\right),\end{aligned}$$

where $\mathbf{M}_1 = \operatorname{diag}\left(\exp(-\boldsymbol{\Sigma}_e(\mathbf{u}_0) - \boldsymbol{\Lambda})\right)$ and $\mathbf{M}_2 = \operatorname{diag}\left(\exp(-\boldsymbol{\Lambda}) - 2\mathbf{I}\right)$.

The ordinary lognormal cokriging estimation variance for each component is

$$\sigma^{o\ell}_{ii}(\mathbf{u}_0) = \alpha_i^2 \exp(\sigma_{ii}(0))\left(1 + \exp(-\sigma^e_{ii}(\mathbf{u}_0) - \lambda_{ii})(\exp(-\lambda_{ii}) - 2)\right). \qquad (5.28)$$

It should be noted that $\vec{\alpha}$ is unknown and therefore only

$$\operatorname{diag}(\vec{\alpha})^{-1}\boldsymbol{\Sigma}_{o\ell}(\mathbf{u}_0)\operatorname{diag}(\vec{\alpha})^{-1} \qquad \text{and} \qquad \frac{\sigma^{o\ell}_{ii}(\mathbf{u}_0)}{\alpha_i\alpha_j}$$

can be computed. The result is only a relative ordinary lognormal cokriging estimation variance, as well as relative lognormal estimation variances for the individual components.

5.3.3 Comments on lognormal cokriging

As presented, lognormal cokriging follows the traditional approach of finding a minimum variance unbiased estimator for the untransformed, original variables. This approach is not always acceptable, as the expected value of a lognormal variate is usually a bad representative of the most frequent value and its variance is large. Taking simple exponentials of the estimated logarithmic values $\ln(\mathbf{X}^*(\mathbf{u}_0))$ would lead to an estimate of the median of the distribution, which is much closer to the mode than the expected value and better represents the more frequent values.

5.4 alr cokriging

The estimation technique known as alr cokriging is used when an r-composition follows a multivariate normal, lognormal, or an additive logistic normal distribution. The latter two cases are particularly concerned with r-compositions; however, because they are built on the former, the case of the normal distribution is presented here first.

5.4.1 alr cokriging with known and unknown mean

Let $\mathbf{Z}(\mathbf{u})$ be an additive logistic normally distributed r-composition with D parts, *i.e.*, $\mathbf{Y}(\mathbf{u}) = \mathrm{alr}(\mathbf{Z}(\mathbf{u}))$ is multivariate normally distributed. The sample space of $\mathbf{Z}(\mathbf{u})$ is the simplex $\mathcal{S}^D$, while the sample space of $\mathbf{Y}(\mathbf{u}) = \mathrm{alr}(\mathbf{Z}(\mathbf{u}))$ is the $(D-1)$-dimensional real space $\mathbb{R}^{D-1}$. Second-order alr stationarity and a partial realization $\{\mathbf{Z}(\mathbf{u}_k), k = 1, 2, \ldots, N\}$ are assumed, and the covariance matrices for every lag $\mathbf{h}$ and for the alr transformed vectors are presumed to be known. The objective is to estimate a vector $\mathbf{Z}(\mathbf{u}_0)$ with an alr linear estimator,

$$\mathbf{Z}^*(\mathbf{u}_0) = \mathrm{agl}\left(\mathbf{c} + \sum_{k=1}^{N} \mathbf{B}_k \mathrm{alr}\left(\mathbf{Z}(\mathbf{u}_k)\right)\right). \tag{5.29}$$

General cokriging makes it possible to obtain

$$\mathbf{Y}^*(\mathbf{u}_0) = \mathbf{c} + \sum_{k=1}^{N} \mathbf{B}_k \mathrm{alr}\left(\mathbf{Z}(\mathbf{u}_k)\right), \tag{5.30}$$

as well as the estimation variance of the alr transformed vectors, both in the case of known and unknown $\mathrm{E}[\mathrm{alr}(\mathbf{Z}(\mathbf{u}))]$. As stated in Aitchison (1986, p. 116), the difficulty of inverse transformation is based on the fact that although first- and second-order moments for an additive logistic normal distribution exist and are finite, they cannot be expressed in an explicit form that allows estimation. It is not possible to express the expected value and the variance for $\mathbf{Z}^*(\mathbf{u}_0)$ in terms of the expected value and variance of $\mathbf{Y}^*(\mathbf{u}_0)$. Thus, an inverse transformation of the estimated value $\mathbf{Y}^*(\mathbf{u}_0)$ that allows an assertion about the properties of $\mathbf{Z}^*(\mathbf{u}_0)$ (about bias, estimation variance, etc.) has not yet been found. Hence, any statement about the properties of the results of estimation can only be applied to alr transformed vectors. Therefore the problem should be formulated in a way that can be answered in terms of logratios, if possible.

Nevertheless, by construction, estimates satisfy the following two conditions: (1) they are estimates of the center of the distribution, defined as $\mathrm{agl}(\mathrm{E}[\mathbf{Y}(\mathbf{u})])$, and are alr unbiased in the sense that the ordinary cokriging estimator $\mathbf{Y}^*(\mathbf{u}_0)$ for $\mathbf{Y}(\mathbf{u}_0)$ at a given location $\mathbf{u}_0$ is unbiased; (2) they minimize the squared Euclidean distance between the alr transformed $\mathbf{Y}(\mathbf{u}_0)$ and its estimator $\mathbf{Y}^*(\mathbf{u}_0)$, as, by construction, the ordinary cokriging estimator minimizes the trace of the covariance matrix of errors. Note that the center—equivalent to the median in univariate lognormal distributions—is a better measure of central tendency for skewed data than is the mean.

If the statement of the problem is formulated in terms of ratios of the original variables, then lognormal cokriging can be applied. It should be remembered that in the case of an unknown mean, the estimation variance will be relative, as it will be expressed in terms of the expected values of the ratios.

As there is no explicit equation which relates $\mathrm{E}[\mathbf{Z}(\mathbf{u})]$ and $\mathrm{E}[\mathrm{alr}(\mathbf{Z}(\mathbf{u}))]$, even when the vector of expected values of the r-composition $\mathbf{Z}(\mathbf{u})$ is known, only cokriging with unknown mean can be applied to the alr transformed vectors. The use of normal cokriging with known mean is possible if the vector of expected values of $\mathrm{alr}(\mathbf{Z}(\mathbf{u}))$ is also known.

If the objective is an interpolation of the original data, the agl transformation (see definition 2.10) can be used. This back transformation leads to an exact interpolation in which the interpolated vectors again form an r-composition. The components add to the same constant as the original r-composition. As with kriging, smoothing also results here (see examples in Section 7.2; Subsections 7.3.1 and 7.4.3). The estimator in equation (5.29) obtained as the agl transformation of either the simple or ordinary cokriging estimator of $\mathbf{Y}^*(\mathbf{u}_0)$ is called either the *simple*, or the *ordinary* alr *cokriging estimator*, and is denoted respectively by $\mathbf{Z}^*_{sa}(\mathbf{u}_0)$ or $\mathbf{Z}^*_{oa}(\mathbf{u}_0)$.

5.4.2 alr cokriging of a subvector

If interest lies in a subvector of an r-composition, and if the sum of its components compared to the theoretical total sum is relatively *small* (for example, if the components are measured in ppm or ppb), then a simpler possibility arises.

Let t be the (theoretical) total sum of the r-composition (for example, $t =$ one million, if the variables are measured in ppm). Let $q(\mathbf{u})$ be the sum at $\mathbf{u} \in \Omega$ of the actual measured components $(\, Z_1(\mathbf{u}) \quad Z_2(\mathbf{u}) \quad \cdots \quad Z_D(\mathbf{u})\,)'$, and define the remainder as $r(\mathbf{u}) = t - q(\mathbf{u})$. If we complete the r-composition

$$\mathbf{Z}(\mathbf{u}) = (\, Z_1(\mathbf{u}) \quad Z_2(\mathbf{u}) \quad \cdots \quad Z_D(\mathbf{u}) \quad r(\mathbf{u})\,)',$$

adding $r(\mathbf{u})$ to the vector of measured variables, the alr transformation using $r(\mathbf{u})$ as a divisor would lead to extremely small values. Thus, for numerical reasons, a scaled alr transformation is implemented using the total sum t. The result of dividing by the variable $r(\mathbf{u})$, multiplying by t, and afterward taking the logarithm differs in numerical terms only slightly or not at all from simply taking the logarithm. In other words, if $Z_i(\mathbf{u})$ is a component of the r-composition $\mathbf{Z}(\mathbf{u})$, then

$$\ln\left(\frac{Z_i(\mathbf{u})}{r(\mathbf{u})}t\right) = \ln(Z_i(\mathbf{u})) + \ln\left(\frac{t}{r(\mathbf{u})}\right);$$

however, under the above conditions,

$$\ln\left(\frac{t}{r(\mathbf{u})}\right) \approx 0 \tag{5.31}$$

and can be ignored. Consequently, with a subvector of this type, the alr transformation can be replaced simply by taking the natural logarithm. Lognormal cokriging can be applied as well without fear of distorting the results. It should be emphasized that condition (5.31) is indispensable.

5.4.3 alr autocorrelation and alr cokriging

From the cokriging system of equations for ordinary cokriging with unknown mean (see property 5.5), it is obvious that for the estimation of $\mathbf{Y}(\mathbf{u}_0)$, the alr cross-covariance matrices enter the system of equations. If the r-composition $\mathbf{Z}(\mathbf{u})$ has null lr autocorrelation for all $h \neq 0$ (definition 4.3) and the spatial covariance structure is symmetric, then property 4.9 yields:

$$\mathbf{T}(\mathbf{h}) = 0 \iff \boldsymbol{\Sigma}(\mathbf{h}) = 0 \quad \forall \mathbf{h} \neq 0. \tag{5.32}$$

The ordinary cokriging system of equations (see property 5.5) can then be written

$$\begin{aligned} \mathbf{B}_\ell \boldsymbol{\Sigma}(0) + \boldsymbol{\Lambda} &= \boldsymbol{\Sigma}(\mathbf{u}_\ell - \mathbf{u}_0), \quad \forall \ell = 1, 2, \ldots, N, \\ \sum_{k=1}^{N} \mathbf{B}_k &= \mathbf{I}, \end{aligned}$$

if $\mathbf{u}_\ell \neq \mathbf{u}_0$. Using equation (5.32), the first equation can be reduced for all $\ell = 1, 2, \ldots, N$ to $\mathbf{B}_\ell \boldsymbol{\Sigma}(0) + \boldsymbol{\Lambda} = 0$; and finally, summing over ℓ, to $\boldsymbol{\Lambda} = -N^{-1}\boldsymbol{\Sigma}(0)$. Then $\mathbf{B}_\ell = N^{-1}\mathbf{I}$ holds for all $\ell = 1, 2, \ldots, N$. This result can be summarized as follows:

Property 5.12 *If an r-composition has null* lr *autocorrelation for all* $\mathbf{h}$ *and the spatial covariance structure is symmetric, then for* $\mathbf{u}_\ell \neq \mathbf{u}_0$, $\ell = 1, 2, \ldots, N$, *ordinary cokriging is equal to taking the average of the transformed vectors.*

Note that the existence of a covariance or correlation in the classic sense (that is, for $\mathbf{h} = 0$) is not ruled out.

By an analogous argument, we obtain the following result for r-compositions with absence of alr cross-correlation (see definition 4.4).

Property 5.13 *If* $\mathbf{Z}(\mathbf{u})$ *has null* alr *cross-correlation for every* $\mathbf{h}$*, then ordinary cokriging of* $\text{alr}(\mathbf{Z}(\mathbf{u}_0))$ *and ordinary kriging of every separate component of* $\text{alr}(\mathbf{Z}(\mathbf{u}_0))$ *are equivalent.*

5.4.4 Permutation invariance of alr cokriging estimators

In preceding chapters it was argued that the use of alr cross-covariances introduced in definition 3.4 is the best way to specify the spatial structure of an r-composition. We advocate this method because the principal objection against it—the nonsymmetry in the components—is not relevant, due to the fact that both the simple and the ordinary alr cokriging estimators are invariant under permutation of the components of the r-composition $\mathbf{Z}(\mathbf{u})$. The proof follows.

Proof. Let $\mathbf{Z}(\mathbf{u})$ be an r-composition, $\mathbf{Y}(\mathbf{u}) = \text{alr}(\mathbf{Z}(\mathbf{u}))$ and let $\mathbf{P}$ be a permutation matrix, $\mathbf{Z}_p(\mathbf{u}) = \mathbf{PZ}(\mathbf{u})$. Further, take $\mathbf{Y}_p(\mathbf{u}) = \text{alr}(\mathbf{Z}_p(\mathbf{u})) = \mathbf{QY}(\mathbf{u})$, where $\mathbf{Q} = \mathbf{FPF}'\mathbf{H}^{-1}$ is a square, nonsingular matrix, with $\mathbf{F}$ as in definition 2.10 and $\mathbf{H}$ as in equation (3.9). The concept of invariance under permutation is illustrated by the diagram shown in Figure 5.1.

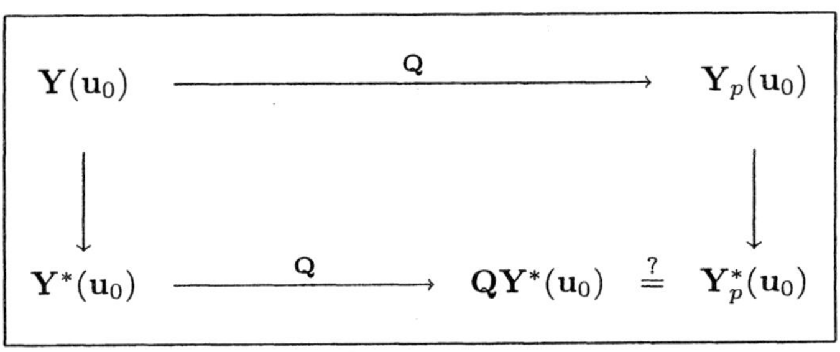

Figure 5.1: Schematic illustration of the concept of invariance under permutation.

We need to prove that $\mathbf{QY}^*(\mathbf{u}_0) = \mathbf{Y}_p^*(\mathbf{u}_0)$, where $\mathbf{Y}^*(\mathbf{u}_0)$ is the ordinary cokriging estimator for $\mathbf{Y}(\mathbf{u}_0)$ and $\mathbf{Y}_p^*(\mathbf{u}_0)$ is the ordinary cokriging estimator for $\mathbf{Y}_p(\mathbf{u}_0)$. The same proof is valid for the corresponding simple cokriging estimators. Let $\mathbf{Y}_p^*(\mathbf{u}_0) = \sum_{k=1}^N \mathbf{B}_k \mathbf{Y}_p(\mathbf{u}_k)$, and $\mathbf{Y}^*(\mathbf{u}_0) = \sum_{k=1}^N \mathbf{A}_k \mathbf{Y}(\mathbf{u}_k)$. Then $\mathbf{QY}^*(\mathbf{u}_0) = \sum_{k=1}^N \mathbf{QA}_k\mathbf{Q}^{-1}\mathbf{Y}_p(\mathbf{u}_k)$, and

$$\begin{aligned}
\mathbf{Y}_p^*(\mathbf{u}_0) = \mathbf{QY}^*(\mathbf{u}_0) \quad &\Longleftrightarrow \quad \sum_{k=1}^N \mathbf{B}_k \mathbf{Y}_p(\mathbf{u}_k) = \sum_{k=1}^N \mathbf{QA}_k\mathbf{Q}^{-1}\mathbf{Y}_p(\mathbf{u}_k) \\
&\Longleftrightarrow \quad \sum_{k=1}^N (\mathbf{B}_k - \mathbf{QA}_k\mathbf{Q}^{-1})\mathbf{Y}_p(\mathbf{u}_k) = 0.
\end{aligned}$$

A sufficient condition for this equation to be satisfied is $\mathbf{B}_k = \mathbf{Q}\mathbf{A}_k\mathbf{Q}^{-1}$. However,

$$\sum_{k=1}^{N} \mathbf{Q}\mathbf{A}_k\mathbf{Q}^{-1} = \mathbf{Q}(\sum_{k=1}^{N} \mathbf{A}_k)\mathbf{Q}^{-1} = \mathbf{Q}\mathbf{Q}^{-1} = \mathbf{I},$$

and the $\mathbf{Q}\mathbf{A}_k\mathbf{Q}^{-1}$ satisfy the same system of equations as the $\mathbf{B}_k$. The solution being unique, $\mathbf{B}_k = \mathbf{Q}\mathbf{A}_k\mathbf{Q}^{-1}$ is obtained for all k. The proof is complete. □

The definitions of the matrix $\mathbf{Q}$ and the agl transformation show that

$$\text{agl}\left(\hat{\mathbf{Y}}_p^*(\mathbf{u}_0)\right) = \mathbf{P}\text{agl}\left(\hat{\mathbf{Y}}^*(\mathbf{u}_0)\right),$$

where $\hat{\mathbf{Y}}_p^*(\mathbf{u}_0) = (\,\mathbf{Y}_p^*(\mathbf{u}_0) \quad 0\,)^t$ and $\hat{\mathbf{Y}}^*(\mathbf{u}_0) = (\,\mathbf{Y}^*(\mathbf{u}_0) \quad 0\,)^t$. [See equation 2.2.] This is obvious from Figure 5.2, which dispenses with representation of the auxiliary vectors $\hat{\mathbf{Y}}_p^*(\mathbf{u})$ and $\hat{\mathbf{Y}}^*(\mathbf{u})$. It follows that not only the cokriging estimators of $\mathbf{Y}(\mathbf{u}_0)$, but also the agl cokriging estimators are invariant under permutation of the components of $\mathbf{Z}(\mathbf{u})$.

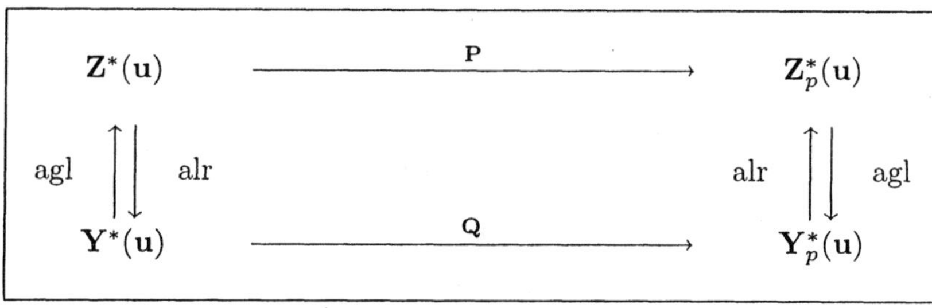

Figure 5.2: Schematic illustration of relationships between estimators of original compositions, alr transformed compositions, and the corresponding permutations.

5.5 Intrinsic vector random functions

Until now, the vector random function $\mathbf{Z}(\mathbf{u})$ has been assumed to be second-order stationary, second-order log-stationary, or second-order alr stationary. The covariance matrices have not been assumed to be symmetric. Computer processing without the assumption of symmetry would probably be only slightly more costly for the modeling and evaluation of the functions describing the spatial covariances and the estimation of coregionalizations. The essential problems lie in the increasing complexity of the modeling process, in the lack of adequate model functions, and in the difficulty of interpretation. In practice, whether or not the r-composition is symmetric in a restricted range of influence (local symmetry) should be investigated. If the r-composition is symmetric, cokriging can be applied within this restricted range of influence. The condition of symmetry for the covariance matrices must also be satisfied—see Matheron (1965) or Myers (1982) for reference—in order to apply the procedure in the second-order stationary case using semivariogram functions, as can be deduced from the fact that for the derivation of the cokriging system of equations, even in the intrinsic case, the following condition of symmetry must be fulfilled:

$$\mathrm{E}[(\mathbf{Z}(\mathbf{u}+\mathbf{h})-\mathbf{Z}(0))(\mathbf{Z}(\mathbf{u})-\mathbf{Z}(0))'] = \mathrm{E}[(\mathbf{Z}(\mathbf{u})-\mathbf{Z}(0))(\mathbf{Z}(\mathbf{u}+\mathbf{h})-\mathbf{Z}(0))'].$$

This condition is necessary in order to write the estimation variance as a function of the matrix of semivariograms and cross-semivariograms. Therefore, in practice, setting up hypotheses of symmetry cannot be avoided. If the condition of symmetry is acceptable, the cokriging system of equations for intrinsic vector random functions can be derived in a manner analogous to the second-order stationary case. This entails a straightforward formal process of derivation and is not given here. A derivation without distributional assumptions was given by Myers (1982).

If $\mathbf{\Gamma}(\mathbf{u}_k-\mathbf{u}_\ell)$ is the matrix of semivariograms and cross-semivariograms (see definition 2.13 for the identity f) of an intrinsic vector random function $\mathbf{Z}(\mathbf{u})$, then the results can be summarized as follows.

1. Cokriging of an intrinsic vector random function $\mathbf{Z}(\mathbf{u})$ with known mean produces

   (a) the simple cokriging estimator: $\mathbf{Z}_s^*(\mathbf{u}_0) = \vec{\mu} + \sum_{k=1}^{N} \mathbf{B}_k\left(\mathbf{Z}(\mathbf{u}_k) - \vec{\mu}\right);$

   (b) a system of equations for simple cokriging: $\sum_{k=1}^{N} \mathbf{B}_k \mathbf{\Gamma}(\mathbf{u}_\ell - \mathbf{u}_k) = \mathbf{\Gamma}(\mathbf{u}_\ell - \mathbf{u}_0), \quad \ell = 1, 2, \ldots, N;$

   (c) the simple cokriging estimation variance $\sigma_s^2 = \sum_{k=1}^{N} \mathrm{Tr}\left(\mathbf{B}_k\mathbf{\Gamma}(\mathbf{u}_0 - \mathbf{u}_k)\right).$

2. Cokriging of an intrinsic vector random function $\mathbf{Z}(\mathbf{u})$ with unknown mean results in:

   (a) an ordinary cokriging estimator: $\mathbf{Z}_o^*(\mathbf{u}_0) = \sum_{k=1}^{N} \mathbf{B}_k\mathbf{Z}(\mathbf{u}_k);$

(b) a system of equations for ordinary cokriging:

$$\sum_{k=1}^{N} \mathbf{B}_k \mathbf{\Gamma}(\mathbf{u}_\ell - \mathbf{u}_k) + \mathbf{\Lambda} = \mathbf{\Gamma}(\mathbf{u}_\ell - \mathbf{u}_0), \qquad \ell = 1, 2, \ldots, N;$$

$$\sum_{k=1}^{N} \mathbf{B}_k = \mathbf{I};$$

(c) an ordinary cokriging estimation variance:

$$\sigma_o^2 = \sum_{k=1}^{N} \mathrm{Tr}\left(\mathbf{B}_k \mathbf{\Gamma}(\mathbf{u}_0 - \mathbf{u}_k)\right) - \mathrm{Tr}(\mathbf{\Lambda}).$$

Analogous equations are obtained for lognormal cokriging.

If we have an alr intrinsic vector random function (see definition 3.10), then the matrix $\mathbf{\Gamma}(\mathbf{u}_\ell - \mathbf{u}_k)$ is replaced by the intrinsic alr cross-covariance matrix $\mathbf{\Psi}(\mathbf{u}_\ell - \mathbf{u}_k)$.

Chapter 6

Practical aspects of compositional data analysis

In Chapter 6 we introduce additional aspects of the geostatistical approach presented in the preceding chapters that were not necessary for its theoretical development, but that are essential for the practical application of the method to compositional data. We discuss how to treat zeros in compositional data sets; how to model the required cross-covariances; how to compute expected values and estimation variances for the original, constrained variables; and how to build and interpret confidence intervals for estimated values.

6.1 Dealing with zeros in compositional data

As mentioned in Section 2.1, data sets with many zeros are as troublesome in compositional analysis as they are in standard multivariate analysis. In our approach, the additional restriction for compositional data is that zero values are not admissible for modeling. The justification for this restriction can be given using arithmetic arguments. A transformation that uses logarithms cannot be performed on zero values. This is the case for the logratio transformation that leads to the definition of an additive logistic normal distribution, as introduced by Aitchison (1986, p. 113). It is also the case for the additive logistic skew-normal distribution defined in Mateu-Figueras et al. (1998), following previous results by Azzalini and Dalla Valle (1996). The centered logratio transformation and the family of multivariate Box-Cox transformations discussed in Andrews et al. (1971), Rayens and Srinivasan (1991), and Barceló-Vidal (1996) also call for the restriction of zero values.

This restriction is certainly a wellspring of discussion, albeit surprisingly so, as nobody would complain about eliminating zeros either by simple suppression of samples or by substitution with reasonable values when dealing with a sample from a lognormal distribution in the univariate case. Recall that the logarithm of zero is undefined and the sample space of the lognormal distribution is the positive real line, excluding the origin.

In order to present our position on how to deal with zeros as clearly as possible, let us assume that only one of our components has zeros in some of the samples. Those cases where more than one variable is affected can be analyzed by methods described below. We distinguish between the following situations:

1. *When zeros correspond to absence of the variable.* As stated in Aitchison (1989, p. 789), "The difference between zero and something is as substantial as the difference between finite and infinite!" Therefore, in this case, at least two populations are present, one defined on the simplex $\mathcal{S}^D$, the other defined on the simplex $\mathcal{S}^{D-1}$, and the data set should be divided into subsets according to presence–absence of the variable.

2. *When zeros correspond to values below detection limit.* Zeros should be substituted according to different trace replacement criteria, like those given by Martín-Fernández et al. (2000) or, when results could easily be distorted by the zeros, they should be amalgamated with other variables, as argued by Martín-Fernández et al. (1997). A possible alternative whose adequacy for regionalized compositions has not been checked is the ranking of data across both cases and variables as proposed by Bacon-Shone (1992). Another approach that might offer a feasible solution is the generalization of replacement methods devised for censored data to spatially dependent data. The latter methods are based on the likelihood function of the sample and require independence. For an introduction to these methods see Sanford et al. (1993) and Buccianti and Rosso (1999).

3. *When zeros correspond to missing values.* This means that the part corresponding to the variable with missing values is actually subsumed in the nonmeasured residual; otherwise the sample would not add to one (or to the corresponding constant) and the missing value could simply be computed by the difference. An amalgamation of the variable with the missing values and the nonmeasured residual following Martín-Fernández et al. (1997) is recommended.

When several variables are affected by the presence of zeros, much effort will be needed to determine which of the possible approaches is appropriate. A different approach might be required for each component or it might even be necessary to split the samples with zeros into different categories of zeros. Thus, no automatic procedure on how to deal with zeros can be expected, as it will always be strongly dependent on the phenomenon under study.

6.2 Modeling alr cross-covariance matrices

It has been argued in previous sections that lr semivariograms or intrinsic variation matrices $\mathbf{\Gamma}(\mathbf{h})$ as introduced in definition 3.9 are the best tools for intrinsic spatial-structure analysis of r-compositions. This conclusion is surprising because, after alr transformation, the obvious counterpart to the usual semivariogram and cross-semivariogram matrix would be the intrinsic alr cross-covariance matrix $\mathbf{\Psi}(\mathbf{h})$ given in definition 3.10. In fact, it is possible and valid to use alr transformed data with standard techniques for vector random functions to estimate and model the intrinsic alr cross-covariance matrix $\mathbf{\Psi}(\mathbf{h})$, given that after alr transformation the constant-sum constraint is suppressed.

Nevertheless, by property 3.15, if either the intrinsic variation matrix $\mathbf{\Gamma}(\mathbf{h})$ or the intrinsic alr cross-covariance matrix $\mathbf{\Psi}(\mathbf{h})$ is known, then the other matrix can be computed directly. This relationship is valid both for theoretical matrix-valued functions and for experimental matrices computed from data applying usual estimators for the semivariograms and cross-semivariograms. Valid models for these matrices should satisfy the same relationships. We will denote the model matrices by $\widehat{\mathbf{\Gamma}}(\mathbf{h})$ and $\widehat{\mathbf{\Psi}}(\mathbf{h})$, respectively, linked by equations established in property 3.15, and determine the conditions that $\widehat{\mathbf{\Gamma}}(\mathbf{h})$ must satisfy in order to be a valid model for $\mathbf{\Gamma}(\mathbf{h})$.

Probably the major difficulty encountered in any process designed to estimate or predict a full coregionalization using cokriging (described in Chapter 5) is devising a positive definite cross-covariance matrix to obtain a valid model. Myers (1982) showed that cross-semivariograms could be expressed in terms of the semivariogram of the pairwise sum of the corresponding variables and the individual semivariograms. If the coregionalization under study has two components, the Cauchy-Schwartz condition on the semivariograms of the pairwise sum of the corresponding variables and the individual semivariograms is a necessary and sufficient condition for the admissibility of models (for admissible simple and cross-semivariograms). This is not true for more than two variables, as stated by Goovaerts (1994a). Kirsch and Pawlowsky (1985) showed that—inasmuch as the Cauchy-Schwartz condition can be expressed either with respect to the cross-semivariogram or the sum semivariogram—in the case of two variables, upper and lower bounds for the region of admissible models for the cross-semivariogram can be obtained.

If the coregionalization under study has D components, $(D-1)^2$ functions must be estimated in the nonsymmetric case. If spatial symmetry is assumed, this reduces to $(D-1)D/2$. Assuming there are enough data to estimate the required functions, the $(D-1)^2$ or $(D-1)D/2$ experimental cross-covariances must be fitted jointly. Until now, the linear model of coregionalization has been used as described by Journel and Huijbregts (1978), Goovaerts (1997), and Olea (1999). This model requires each autocovariance $C_{kk}(h)$ and each cross-covariance $C_{kk'}(h)$ to be a different linear combination of the same L basic permissible covariance functions $C_\ell(h)$, *i.e.*

$$C_{kk'}(h) = \sum_{\ell=1}^{L} b_{kk'}^{(\ell)} C_\ell(h),$$

for all k, k'.

Positive definiteness of the cross-covariance matrix is then ensured as long as the L matrices of coefficients $[b_{kk'}^{(\ell)}]$, $\ell = 1, 2, ..., L$ are themselves positive definite. As indicated by Journel (1998), this model is sufficient but not necessary to ensure positive definiteness. Its main limitation is the necessity to fit all of the necessary functions by the same set of basic functions $C_\ell(h)$.

Goulard and Voltz (1992) proposed a semiautomatic algorithm to derive the matrices of coefficients $[b_{kk'}^{(\ell)}]$, but it can only be used after the basic components have been determined. Bourgault and Marcotte (1991) present the "multivariable variogram." Combined with the linear model of coregionalization, it provides a way to find the elementary semivariograms that characterize different spatial scales. When the number of elementary components is less than or equal to the number of variables, it is possible by means of linear regression of semivariograms and cross-variograms to estimate the coregionalization parameters directly. In this manner the elementary variables themselves can be obtained, either by cokriging or by direct matrix inversion. The problem with the approach actually lies in the estimation of the population covariance matrix from (spatially) correlated samples, which is not addressed by Bourgault and Marcotte (1991). An interesting exposition of problems related to covariance estimation in the univariate case was published by Barnes (1991); no similar discussion for the multivariate case is known to the authors, although it is to be expected that results can be extended directly.

Yao and Journel (1998) and Ma and Yao (2001) have implemented a completely different approach to autocovariance and cross-covariance matrix modeling using spectral properties of the corresponding autocorrelation and cross-correlation matrices. For D variables, the $D(D-1)/2$ experimental cross-correlation functions are written as $N^2D(D-1)/2$ discrete tables, each in $\mathbb{R}^2$, where N^2 is the number of points into which the actual auto- and cross-correlation functions have been discretized. After estimating the auto- and cross-correlation values at all possible points, missing values are filled in by allowing for a larger tolerance; the values are then smoothed and finally FFT (Fast Fourier Transform) is applied to obtain "quasi" spectrum tables. Joint positive definiteness is easily achieved in the transformed space by perturbing, smoothing, and standardizing the spectrum frequency values. The result is back transformed into jointly positive definite auto- and cross-correlation tables. Multiplication by experimental variance and covariance estimates leads us finally to jointly positive definite auto- and cross-covariance tables, from which all required values are read directly. No assumption of symmetry or isotropy is required for this approach. The process is far from fully automated and still requires a great deal of expertise. This is evident from the following short description of the practical implementation procedure, involving three utility programs: one for steps 2 and 3, another for step 4, and a third one for steps 5 and 6.

Step 1. Define a two-dimensional neighborhood map $\mathcal{M}$, or maximum search window, that will be used for estimation. Discretize it in such a way that the foreseeable directional distances $\mathbf{u}_i - \mathbf{u}_0$ between sampling points $\mathbf{u}_i$ and points $\mathbf{u}_0$, where the estimation is required, are reasonably approximated by one node. We suggest an internodal distance one order of magnitude smaller than the average distance between observation sites. Half the maximum extension of the map is equivalent to the largest lag for which a univariate autocorrelogram or semivariogram is usually computed and should be smaller than half the extension of the domain of interest Ω. The implementation of the software corresponding to the method of Ma and Yao (2001) requires that the number of columns and rows be necessarily $(1 + 2^k)$, where k is an integer.

Step 2. Assign every vectorial distance $\mathbf{u}_i - \mathbf{u}_j$, where $\mathbf{u}_i$ and $\mathbf{u}_j$ correspond to sampling points inside Ω, to the closest node in $\mathcal{M}$. Once all possible combinations of pairs have been exhausted, all D autocorrelograms and $D(D-1)/2$ cross-correlograms are calculated point by point with the available pairs of data, if any. To reduce erratic fluctuations, the user could eliminate all those correlogram values with less than a certain number of pairs, a decision that should relate closely to the spatial continuity of the data and the sample size. In general the less continuous the variables and the larger the sample size, the more sense it makes to set a high threshold. This first set of correlogram maps is used only as a reference to compare results from further processing. In this and following steps, displaying the auto- and cross-correlation neighborhood maps in color and preparing, in addition, the more conventional cross-sectional displays in at least four directions makes the evaluation of results easier.

Step 3. Use moving windows to estimate a new set of auto- and cross–correlogram values for every node. Although the window size can be any arbitrary number of columns and rows around every node, actual practice indicates that one or two columns and rows should be adequate. For example, disregarding border effects, a halo of two nodes implies that all pairs in a 5-by-5 array of nodes centered at every node enter into the calculation of the auto- and cross-correlograms for that node. Because of overlapping, the process produces smoothing of the maps which the user could increase even further with a threshold that disregards estimations with less than a minimum number of pairs of points. Notice that this averaging is different than the traditional way of estimating an experimental correlogram or a semivariogram a single direction at a time, in which pairs are grouped into classes that do not overlap.

Step 4. Use interpolation as a final step to guarantee that 100% of the nodes have an associated correlation value. Although maps resulting from step 2 commonly have nodes without a correlogram value, their total number can be significantly reduced using a halo of $\pm$ one or two columns and rows, as described in step 3.

Step 5. Force the auto- and cross-correlogram maps to fulfill the conditions for real (not imaginary) spectral densities. The maps resulting from step 4 are projected into the spectral domain using an FFT but are not necessarily true spectral densities. They must be jointly perturbed in order to fulfill all necessary conditions. The process is controlled by windows, as described in the next step.

Step 6. Initialize the starting and stopping of the perturbation (actually "smoothing") procedure. An initial and a final window are required. The solution is not unique, as can be seen from the fact that results, in general, change with increasing sizes of windows. The larger the window, the smoother the result. In the (usual) absence of stopping criteria, the most sensible course of action is to emulate the practice of manual modeling of semivariograms: Make a graphical comparison of back transformed spectral maps with the initial sample auto- and cross-correlation maps described in step 2. The window-size combination producing the closest fit to the original sample maps seems to be the model of choice, although more consistent criteria should be expected from a more extensive experience with this kind of modeling procedure. In the process we have described, the maps lose the outermost halo of rows and columns, which means the final number of rows and columns is $(2^k - 1)$, where k is an integer.

Step 7. Rescale the resulting maps, which are permissible auto- and cross–correlogram models. After rescaling with the appropriate estimates of variances and covariances, transform the maps into auto- and cross-covariance lookup tables.

Employing the method of Ma and Yao (2001), processing is rapid, as every map series is generated by a single execution of one of the utility programs.

6.3 Exploratory analysis of compositional data

The most common concepts for exploratory analysis of any type of data without taking into account their spatial location are the arithmetic mean and median as measures of central tendency; the sample variance and standard deviation as measures of dispersion; minimum, maximum, quantiles, and range as general descriptive characteristics of the empirical distribution of the data; and the histogram and box-plot as corresponding univariate graphical representations of the sample. Aitchison (1997) argued that the arithmetic mean is not representative of a compositional data set; he suggested, as a more realistic alternative, the closure of the geometric mean as a measure of central tendency or center:

$$
\begin{aligned}
\mathrm{cen}(\mathbf{Z}) &= \check{C}\,(g(Z_1) \quad g(Z_2) \quad \cdots \quad g(Z_D))' && (6.1)\\
&= \frac{1}{g_s}\,(g(Z_1) \quad g(Z_2) \quad \cdots \quad g(Z_D))', && (6.2)
\end{aligned}
$$

where $g(Z_i)$ is the geometric mean of the ith component as given in definition 2.11 and $g_s = g(Z_1) + g(Z_2) + \cdots + g(Z_D)$. Arguments are based on minimization of the Kullback-Leibler directed divergence as described in Aitchison (1997). A more intuitive argument can be based on an analogy to the lognormal distribution where the arithmetic mean is also an unsuitable measure of central tendency and therefore is frequently replaced by the median, or even the mode. Note that in the case of compositional data, the vector of medians and the vector of modes need not necessarily be a composition before closure (as can be seen in Table 7.1, in an application to real data).

In a similar manner, neither the sample variance nor the standard deviation can be considered as an adequate measure of dispersion, especially when the problem is considered principally from a multivariate point of view, which is actually the only sensible perspective for compositional data. A measure of dispersion is important for comparisons between different data sets and therefore a measure of total variability as a measure of dispersion around the center of the data makes sense. Aitchison (1997) defines such a measure as

$$\mathrm{Totvar}(\mathbf{Z}) = \frac{1}{D}\sum_{i<j} \mathrm{Var}\left[\ln\left(\frac{Z_i}{Z_j}\right)\right]. \tag{6.3}$$

In the practical applications in Chapter 7, all of these descriptive statistics are used. Although we agree with the arguments against the arithmetic mean and the variance or standard variation, we think it is useful to compare their descriptive power.

See Martín-Fernández et al. (1998a) for a discussion related to other measures of difference for compositional data sets; Aitchison (1997) describes other exploratory tools such as the biplot.

6.4 Back transforming alr means and variances

Once the estimation procedure has been applied to the alr transformed data set and estimations in $\mathbb{R}^{D-1}$ have been obtained, interest may lie in deriving estimates in the original sample space (i.e., in the simplex $\mathcal{S}^D$), along with an estimation variance or prediction error. As mentioned in Subsection 5.4.1, there is no explicit form of the integral required to obtain estimates of the expected value or the variance. Nonetheless, it is possible to use multivariate Gauss-Hermite quadrature to get a reasonable approximation of the desired integrals.

Properties of matrices and determinants which can be found in many standard textbooks such as the one by Golub and Van Loan (1989) are required in the following. The purpose is to compute an estimate of

$$\vec{\mu}_{\mathbf{Z}} = \mathrm{E}[\mathbf{Z}] = \int_{\mathcal{S}^D} \mathbf{Z} f(\mathbf{Z}) d\mathbf{Z} \tag{6.4}$$

and

$$\boldsymbol{\Sigma}_{\mathbf{Z}} = \mathrm{Cov}[\mathbf{Z}, \mathbf{Z}] = \int_{\mathcal{S}^D} (\mathbf{Z} - \vec{\mu}_{\mathbf{Z}})(\mathbf{Z} - \vec{\mu}_{\mathbf{Z}})' f(\mathbf{Z}) d\mathbf{Z}\,, \tag{6.5}$$

where

$$f(\mathbf{Z}) = (2\pi)^{-\frac{d}{2}} |\boldsymbol{\Sigma}_{\mathbf{X}}|^{-\frac{1}{2}} \left(\prod_{i=1}^{D} Z_i\right)^{-1} \exp\left(-\tfrac{1}{2}(\mathrm{alr}(\mathbf{Z}) - \vec{\mu}_{\mathbf{X}})' \boldsymbol{\Sigma}_{\mathbf{X}}^{-1} (\mathrm{alr}(\mathbf{Z}) - \vec{\mu}_{\mathbf{X}})\right) \tag{6.6}$$

is obtained under the assumption that $\mathbf{X} = \mathrm{alr}(\mathbf{Z})$, and follows a multivariate normal distribution. Recall that if $\mathbf{X}$ follows a multivariate normal distribution, then

$$f(\mathbf{X}) = (2\pi)^{-\frac{d}{2}} |\boldsymbol{\Sigma}_{\mathbf{X}}|^{-\frac{1}{2}} \exp\left(-\tfrac{1}{2}(\mathbf{X} - \vec{\mu}_{\mathbf{X}})' \boldsymbol{\Sigma}_{\mathbf{X}}^{-1} (\mathbf{X} - \vec{\mu}_{\mathbf{X}})\right),$$

and for $\mathbf{X} = \mathrm{alr}(\mathbf{Z})$ and, respectively, $\mathbf{Z} = \mathrm{agl}(\mathbf{X})$, the Jacobian of the transformation is

$$J_{\mathrm{alr}(\mathbf{Z})} = \left(\prod_{i=1}^{D} Z_i\right)^{-1}.$$

Furthermore, recall that the transformation of a random variable leads to the following relationship between the corresponding distribution functions:

$$\begin{aligned} F(\mathbf{Z}_0) &= \mathrm{P}[\mathbf{Z}_0 \in Z_0] = \mathrm{P}[\mathrm{agl}^{-1}(\mathbf{Z}_0) \in \mathrm{agl}^{-1}(Z_0)] = \mathrm{P}[\mathbf{X}_0 \in \mathrm{alr}^{-1}(Z_0)] \\ &= \int_{\mathrm{alr}(Z_0)} f(\mathbf{X}) d\mathbf{X} = \int_{Z_0} J_{\mathrm{alr}(\mathbf{Z})} f(\mathrm{alr}(\mathbf{Z})) d\mathbf{Z}\,. \end{aligned}$$

In terms of cokriging, $\vec{\mu}_{\mathbf{Z}}$ in equations (6.4), (6.5), and (6.6) will be, in practice, the vector of predicted values and $\boldsymbol{\Sigma}_{\mathbf{Z}}$ will be the cokriging error variance matrix in $\mathcal{S}^D$.

As mentioned by Aitchison (1986, p. 314), multivariate Gauss-Hermite integration of order k can be used to obtain the approximation

$$\int_{\mathbb{R}^D} g(\mathbf{Y}) \exp(-\mathbf{Y}'\mathbf{Y}) d\mathbf{Y} \approx \sum_{i_1=1}^{k} \sum_{i_2=1}^{k} \cdots \sum_{i_D=1}^{k} w_{i_1} w_{i_2} \cdots w_{i_D} g(Y_{i_1}, Y_{i_2}, \ldots, Y_{i_D}), \tag{6.7}$$

which can be computed either directly or by using the following multidimensional product integration rule (see, e.g., Davis and Rabinowitz 1975, pp. 268–275):

$$\approx \sum_{i_1=1}^{k} w_{i_1} \sum_{i_2=1}^{k} w_{i_2} \cdots \sum_{i_D=1}^{k} w_{i_D} g(Y_{i_1}, Y_{i_2}, \cdots, Y_{i_D}),$$

for known weights w_1, w_2, …, w_k and known abscissae Y_1, Y_2, …, Y_k.

To verify that Gauss-Hermite integration can be used, we need to express equations (6.4) and (6.5) as

$$\vec{\mu}_{\mathbf{Z}} = \int_{\mathbb{R}^D} g_1(\mathbf{Y}) \exp(-\mathbf{Y}'\mathbf{Y}) d\mathbf{Y} \tag{6.8}$$

and

$$\mathbf{\Sigma_Z} = \int_{\mathbb{R}^D} g_2(\mathbf{Y}) \exp(-\mathbf{Y}'\mathbf{Y}) d\mathbf{Y}. \tag{6.9}$$

To obtain these expressions, we take

$$\mathbf{Y}'\mathbf{Y} = \tfrac{1}{2}(\mathrm{alr}(\mathbf{Z}) - \vec{\mu}_{\mathbf{X}})'\mathbf{\Sigma}_{\mathbf{X}}^{-1}(\mathrm{alr}(\mathbf{Z}) - \vec{\mu}_{\mathbf{X}}).$$

Given that $\mathbf{\Sigma_X}$ is square, symmetric and positive-definite (or positive–semi-definite), we can obtain its Cholesky decomposition. An upper triangular matrix R is obtained, which satisfies the following conditions:

$$\begin{aligned}
\mathbf{\Sigma_X} &= R'R; \\
\mathbf{\Sigma}_{\mathbf{X}}^{-1} &= (R'R)^{-1} = R^{-1}(R')^{-1} = R^{-1}\left(R^{-1}\right)'.
\end{aligned}$$

Consequently,

$$\mathbf{Y}'\mathbf{Y} = \tfrac{1}{\sqrt{2}}(\mathrm{alr}(\mathbf{Z}) - \vec{\mu}_{\mathbf{X}})'R^{-1}\tfrac{1}{\sqrt{2}}\left(R^{-1}\right)'(\mathrm{alr}(\mathbf{Z}) - \vec{\mu}_{\mathbf{X}}), \tag{6.10}$$

and finally

$$\mathbf{Y} = \tfrac{1}{\sqrt{2}}\left(R^{-1}\right)'(\mathrm{alr}(\mathbf{Z}) - \vec{\mu}_{\mathbf{X}}). \tag{6.11}$$

So, $\mathbf{Z}$ can be computed immediately for known $\mathbf{Y}$. In fact, from equation (6.11), the following sequence of identities is easily derived:

$$\begin{aligned}
\sqrt{2}\mathbf{Y} &= \left(R^{-1}\right)'(\mathrm{alr}(\mathbf{Z}) - \vec{\mu}_{\mathbf{X}}) \\
\sqrt{2}R'\mathbf{Y} &= \mathrm{alr}(\mathbf{Z}) - \vec{\mu}_{\mathbf{X}} \\
\sqrt{2}R'\mathbf{Y} + \vec{\mu}_{\mathbf{X}} &= \mathrm{alr}(\mathbf{Z}).
\end{aligned}$$

Using now the inverse of the alr transformation, we get

$$\mathbf{Z} = \mathrm{agl}\left(\sqrt{2}R'\mathbf{Y} + \vec{\mu}_{\mathbf{X}}\right). \tag{6.12}$$

Using equation (6.12), taking into account (6.10) and (6.11), and substituting in (6.6), it is seen that the expressions in (6.8) and (6.9) can be obtained readily. But before computing the required functions $g_1(.)$ and $g_2(.)$, we need the Jacobian of the transformation.

Given equation (6.12), if we write $\mathbf{X} = \sqrt{2}R'\mathbf{Y} + \vec{\mu}_\mathbf{X}$, we obtain $\mathbf{Z} = \mathrm{agl}(\mathbf{X})$ and

$$\left|\frac{\partial \mathbf{X}}{\partial \mathbf{Z}}\right| = \left(\prod_{i=1}^{D} Z_i\right)^{-1},$$

as stated by Aitchison (1986, p. 115). Furthermore, given that $\vec{\mu}_\mathbf{X}$ is a constant with respect to $\mathbf{Y}$ and that R' is lower triangular,

$$\left|\frac{\partial \mathbf{X}}{\partial \mathbf{Y}}\right| = \left|\frac{\partial\left(\sqrt{2}R'\mathbf{Y} + \vec{\mu}_\mathbf{X}\right)}{\partial \mathbf{Y}}\right| = \left|\frac{\partial\left(\sqrt{2}R'\mathbf{Y}\right)}{\partial \mathbf{Y}}\right| = \sqrt{2}^{D-1}\left|R'\right| = 2^{\frac{D-1}{2}}\left|R'\right|.$$

R being the Cholesky decomposition of $\mathbf{\Sigma_X}$, it holds that

$$|R'| = |R| = |\mathbf{\Sigma_X}|^{\frac{1}{2}}.$$

Therefore

$$\partial \mathbf{Z} = \prod_{i=1}^{D} Z_i \partial \mathbf{X} = \prod_{i=1}^{D} Z_i 2^{\frac{D-1}{2}} |R|\, \partial \mathbf{Y} = \prod_{i=1}^{D} Z_i 2^{\frac{D-1}{2}} |\mathbf{\Sigma_X}|^{\frac{1}{2}}\, \partial \mathbf{Y}, \tag{6.13}$$

and finally,

$$\partial \mathbf{Y} = 2^{-\frac{D-1}{2}}\left(\prod_{i=1}^{D} Z_i\right)^{-1} |\mathbf{\Sigma_X}|^{-\frac{1}{2}}\, \partial \mathbf{Z}.$$

Substituting equations (6.12) and (6.13) in (6.4) and (6.5), we obtain (6.8) and (6.9) as follows:

$$\vec{\mu}_\mathbf{Z} = \int_{\mathbb{R}^{D-1}} \pi^{-\frac{D-1}{2}} \mathrm{agl}\left(\sqrt{2}R'\mathbf{Y} + \vec{\mu}_\mathbf{X}\right) \exp(-\mathbf{Y}'\mathbf{Y}) d\mathbf{Y}$$

and

$$\mathbf{\Sigma_Z} = \int_{\mathbb{R}^{D-1}} \pi^{-\frac{d}{2}} \mathbf{M}\mathbf{M}' \exp(-\mathbf{Y}'\mathbf{Y}) d\mathbf{Y},$$

with $\mathbf{M} = \mathrm{agl}\left(\sqrt{2}R'\mathbf{Y} + \vec{\mu}_\mathbf{X}\right) - \vec{\mu}_\mathbf{Z}$. To compute $\vec{\mu}_\mathbf{Z}$, using equation (6.7), we take

$$g_1 = \pi^{-\frac{D-1}{2}} \mathrm{agl}\left(\sqrt{2}R'\mathbf{Y} + \vec{\mu}_\mathbf{X}\right)$$

and to compute $\mathbf{\Sigma_Z}$ we take

$$g_2 = \pi^{-\frac{D-1}{2}}\left(\mathrm{agl}\left(\sqrt{2}R'\mathbf{Y} + \vec{\mu}_\mathbf{X}\right) - \vec{\mu}_\mathbf{Z}\right)\left(\mathrm{agl}\left(\sqrt{2}R'\mathbf{Y} + \vec{\mu}_\mathbf{X}\right) - \vec{\mu}_\mathbf{Z}\right)'.$$

To see that $J_{\mathrm{alr}(\mathbf{Z})} = \left(\prod_{i=1}^{D} Z_i\right)^{-1}$ holds, consider

$$\begin{aligned}\mathrm{alr}(\mathbf{Z}) &= \begin{pmatrix} \ln \dfrac{Z_1}{Z_D} & \ln \dfrac{Z_2}{Z_D} & \cdots & \ln \dfrac{Z_{D-1}}{Z_D} \end{pmatrix} \\ &= \begin{pmatrix} \ln(Z_1) - \ln(Z_D) & \ln(Z_2) - \ln(Z_D) & \cdots & \ln(Z_{D-1}) - \ln(Z_D) \end{pmatrix};\end{aligned}$$

consequently,

$$J_{\mathrm{alr}(\mathbf{Z})} = \begin{vmatrix} \frac{\partial(\ln(Z_1)-\ln(Z_D))}{\partial Z_1} & \frac{\partial(\ln(Z_2)-\ln(Z_D))}{\partial Z_1} & \cdots & \frac{\partial(\ln(Z_{D-1})-\ln(Z_D))}{\partial Z_1} \\ \frac{\partial(\ln(Z_1)-\ln(Z_D))}{\partial Z_2} & \frac{\partial(\ln(Z_2)-\ln(Z_D))}{\partial Z_2} & \cdots & \frac{\partial(\ln(Z_{D-1})-\ln(Z_D))}{\partial Z_2} \\ \vdots & \vdots & \ddots & \vdots \\ \frac{\partial(\ln(Z_1)-\ln(Z_D))}{\partial Z_D} & \frac{\partial(\ln(Z_2)-\ln(Z_D))}{\partial Z_D} & \cdots & \frac{\partial(\ln(Z_{D-1})-\ln(Z_D))}{\partial Z_D} \end{vmatrix},$$

and, given that $Z_D = \sum_{i=1}^{D-1} Z_i$, the elements on the principal diagonal are all of the form

$$\frac{\partial(\ln(Z_i) - \ln(Z_D))}{\partial Z_i} = \frac{1}{Z_i} + \frac{1}{Z_D},$$

while the off-diagonal elements are all equal:

$$\frac{\partial(\ln(Z_j) - \ln(Z_D))}{\partial Z_i} = 0 + \frac{1}{Z_D}.$$

We then get

$$J_{\mathrm{alr}(\mathbf{Z})} = \begin{vmatrix} \frac{1}{Z_1} + \frac{1}{Z_D} & \frac{1}{Z_D} & \cdots & \frac{1}{Z_D} \\ \frac{1}{Z_D} & \frac{1}{Z_2} + \frac{1}{Z_D} & \cdots & \frac{1}{Z_D} \\ \vdots & \vdots & \ddots & \vdots \\ \frac{1}{Z_D} & \frac{1}{Z_D} & \cdots & \frac{1}{Z_D} + \frac{1}{Z_D} \end{vmatrix}.$$

Multiplying the first column by Z_1, which implies dividing the determinant by the same value, and adding to the first column all the other columns multiplied by the corresponding Z_i, with i the column index, and taking into account that

$$1 + \sum_{i=1}^{D-1} \frac{Z_i}{Z_D} = \frac{1}{Z_D},$$

leads to the following expression for the Jacobian:

$$J_{\mathrm{alr}(\mathbf{Z})} = \frac{1}{Z_1} \begin{vmatrix} \frac{1}{Z_D} & \frac{1}{Z_D} & \cdots & \frac{1}{Z_D} \\ \frac{1}{Z_D} & \frac{1}{Z_2} + \frac{1}{Z_D} & \cdots & \frac{1}{Z_D} \\ \vdots & \vdots & \ddots & \vdots \\ \frac{1}{Z_D} & \frac{1}{Z_D} & \cdots & \frac{1}{Z_D} + \frac{1}{Z_D} \end{vmatrix}.$$

If the first column is now subtracted from each one of the others, the result is

$$J_{\mathrm{alr}(\mathbf{Z})} = \frac{1}{Z_1} \begin{vmatrix} \frac{1}{Z_D} & 0 & \cdots & 0 \\ \frac{1}{Z_D} & \frac{1}{Z_2} & \cdots & 0 \\ \vdots & \vdots & \ddots & \vdots \\ \frac{1}{Z_D} & 0 & \cdots & \frac{1}{Z_D} \end{vmatrix} = \prod_{i=1}^{D} \frac{1}{Z_i} = \left(\prod_{i=1}^{D} Z_i \right)^{-1} .$$

6.5 Confidence intervals and confidence regions

Confidence intervals and confidence regions have been introduced historically as alternatives to point estimates for unknown parameters. The reason for this is that point estimates give no hint of how far or how close the unknown but true value might be from the value obtained as an estimate.

6.5.1 General concepts

Confidence intervals and confidence regions are not only helpful in assessing the accuracy and variability of estimators, but also in making decisions about the unknown parameters. However, confidence intervals and confidence regions are estimates, and should not be confused with probability intervals because, once computed, no random variable or random vector is involved. We may interpret a confidence interval as follows: Suppose several samples of the same size are taken from the same population; suppose further that from each one of those samples, a $100(1-\alpha)\%$ confidence interval for an unknown parameter is constructed; then $100(1-\alpha)\%$ of those confidence intervals will contain the unknown parameter. The value $100(1-\alpha)$ is called the *confidence coefficient*.

In Section 1.1 it was mentioned that interpreting an estimation error as a standard deviation of an estimated component may lead to an erroneous definition of a symmetric confidence interval, by implicitly assuming that the predicted component is at least symmetric. In addition to that, if an approximate confidence is assigned to the interval, usually 95% for a confidence interval centered at the estimated component plus or minus two times the standard deviation of the estimator, then a Gaussian model is assumed. But a Gaussian model requires an unconstrained sample space, in which the variables range, at least theoretically, from minus infinity to plus infinity. As a consequence it is not unusual to get confidence intervals for data measured in percentages or ppm that include negative, and thus impossible, values. For an example, see Goovaerts (1997, p. 261). In a multivariate context, a confidence region in the form of a multidimensional ellipsoid can be defined around the estimated vector, as pointed out by ver Hoef and Cressie (1993). For illustrative purposes, consider three components that form a composition. Valid realizations of the components must satisfy the constant-sum constraint and thus have to be included in the corresponding simplex, which in this case is nothing but a ternary diagram. A three-dimensional ellipsoid centered around a predicted value will include possible realizations only in the intersection of the ternary diagram and the ellipsoid, while those outside the intersection will not satisfy the constant-sum constraint, as illustrated in Figure 6.1.

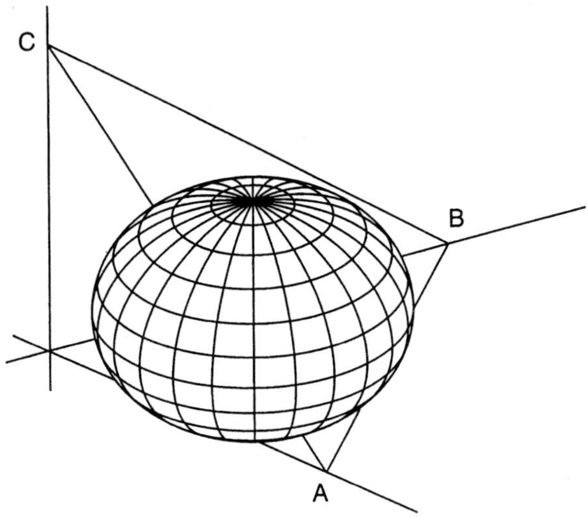

Figure 6.1: Confidence ellipsoid and ternary diagram.

Mood et al. (1986) give an excellent introduction to standard construction of confidence intervals, and Dowd (1989) and Tercan and Dowd (1995) present detailed discussions on the importance of nonsymmetrical confidence intervals in the univariate regionalized case. Krzanowski (1988) is an adequate reference for the concept of confidence regions in general; ver Hoef and Cressie (1993) deal with the subject in the regionalized case. We will follow these steps and present first the classical concepts of confidence interval and confidence region, and proceed afterward to analyze their importance in the case of regionalized compositions.

6.5.2 Confidence intervals and confidence regions in the nonregionalized case

In the framework of classical statistics, consider a random variable X whose density function depends on an unknown parameter θ. A sample $X_1, X_2, ..., X_N$ is defined as N random variables that are identically distributed with X; if the variables are independent, the sample is called a *random sample.* Recall that the requirement of independence is not fulfilled when dealing with regionalized variables, and that the sample is considered as a set of random variables *before* the experiment is performed and as numbers *afterward.*

In its most general formulation, a confidence interval for θ is obtained from a *random sample* as follows:

1. Select a function of the sample $Q = g(X_1, X_2, ..., X_N)$ such that the density function f_Q of Q is completely determined or depends only on θ;

2. for a desired confidence level $1 - \alpha$, find two functions $h_1(\theta)$ and $h_2(\theta)$ such that for any given value of θ,

$$\begin{aligned} \mathrm{P}[h_1(\theta) < Q \leq h_2(\theta)] &= \mathrm{P}[Q \leq h_2(\theta)] - \mathrm{P}[Q \leq h_1(\theta)] \\ &= \int_{h_1(\theta)}^{h_2(\theta)} f_Q(t)dt = 1 - \alpha\,. \end{aligned} \tag{6.14}$$

3. Substitute Q in $[Q \leq h_1(\theta)]$ and in $[Q \leq h_2(\theta)]$ by its realization and transform it in such a way that a lower and an upper bound for θ in the parameter space of θ is obtained; the result is the desired confidence interval.

The functions $h_1(\theta)$ and $h_2(\theta)$ can sometimes be found explicitly by solving the corresponding integral in equation (6.14); if Q depends on θ but f_Q does not, then both limiting functions are constants and can be obtained by direct calculation or from statistical tables. In some cases a numerical integration or more crude graphical methods might be useful to obtain desired results. See Mood and Graybill (1963, pp. 256–260) for a similar presentation and further details.

Although no restrictions have been imposed on the choice of Q up to this point, it is standard practice to use either maximum likelihood or method-of-moments estimators, or a function of one or the other, with the purpose of finding optimal confidence intervals in the sense of minimum length.

To illustrate this method for constructing confidence intervals, which is seldom found in standard statistical textbooks, assume X follows a Gaussian distribution with unknown expected value μ and variance σ^2. If we use the arithmetic mean to estimate $\theta = \mu$, we get $\bar{X} = (1/N)\sum_{n=1}^{N} X_n$ and $\bar{X}$ follows a Gaussian distribution with expected value μ and variance σ^2/N. But the density function of $\bar{X}$ depends on two unknown parameters. Therefore, we use the fact that for a random sample the standardized variable $Q = (\bar{X}-\mu)/(S/\sqrt{N})$, where S is the experimental standard deviation, follows a Student's $\mathtt{t}$ distribution on $N-1$ degrees of freedom. Thus, we can readily obtain $h_1(\mu)$ and $h_2(\mu)$ from the tables of Student's $\mathtt{t}$ distribution for any desired value of α, as in this case both functions are constant. For a symmetric confidence interval, $h_1(\theta) = -\mathtt{t}_{\alpha/2}$ and $h_2(\theta) = \mathtt{t}_{\alpha/2}$, and

$$\mathrm{P}\left[-\mathtt{t}_{\alpha/2} < \frac{\bar{X}-\mu}{S/\sqrt{N}} \leq \mathtt{t}_{\alpha/2}\right] = 1-\alpha$$

or, equivalently,

$$\mathrm{P}\left[\bar{X} - \mathtt{t}_{\alpha/2}S/\sqrt{N} < \mu \leq \bar{X} + \mathtt{t}_{\alpha/2}S/\sqrt{N}\right] = 1-\alpha, \tag{6.15}$$

which is a well-known result.

If X is not Gaussian, then the central limit theorem establishes that if N is large, then $Q = (\bar{X}-\mu)/(S/\sqrt{N})$ is approximately normal with expected value 0 and unit variance. This result is of crucial importance. It guarantees that, in most cases, a symmetric confidence interval for the expected value of a random variable, estimated by the arithmetic mean of the sample, can be obtained by substituting $\mathtt{t}_{\alpha/2}$ in equation (6.15) by the corresponding $\alpha/2$ percentile of the standard normal distribution, $\delta_{\alpha/2}$. But to say *in most cases* does not mean *in all cases*, as the requirement of large N is quite arbitrary; the degree of convergence to the normal distribution depends heavily on the distribution of the parent population, especially when it is skewed. According to Krzanowski (1988), a standard value of 25 is assumed to be reasonable. Recall that for $\alpha = 0.05$, $\delta_{\alpha/2} = 1.96$, thus we obtain the frequently used approximate 95% confidence interval as the arithmetic mean plus or minus two times the experimental standard deviation $s/\sqrt{N}$.

If we assume now that $\mathbf{X}$ is a random vector and not a single random variable, we must define confidence regions. Confidence regions are not just a set of confidence intervals defined separately for each component of the random vector; they take into account the relationship between the components when the components are not independent from one another. Therefore the joint density for a function of the sample is needed to construct a confidence region, while the rest of the approach remains essentially the same. Thus, given a random vector $\mathbf{X}$ whose multivariate density depends on a vector of parameters $\vec{\theta}$ and a random sample $\mathbf{X}_1, \mathbf{X}_2, ..., \mathbf{X}_N$, the following steps are required:

1. selecting a vector function of the sample $Q = g(\mathbf{X}_1, \mathbf{X}_2, ..., \mathbf{X}_N)$ such that the multivariate density function f_Q of Q is completely determined or depends only on the vector $\vec{\theta}$;

2. for a desired confidence level $1 - \alpha$, finding a vector function $h(\vec{\theta})$ such that for any given value of the vector of parameters $\vec{\theta}$,

$$\mathrm{P}[|Q| \leq h(\vec{\theta})] = 1 - \alpha; \tag{6.16}$$

3. substituting Q in $|Q| \leq h(\vec{\theta})$ by its realization and transforming it in such a way that a region in the parameter space of $\vec{\theta}$ is defined; the result is the desired confidence region.

Although this presentation is analogous to the construction presented for the univariate case, it is usually not as easy to construct confidence regions as it is to construct confidence intervals. As mentioned by Mood and Graybill (1963, p. 258), a graphical approach is excluded in the construction process. Note that the procedure is always specific to the distribution function used, unless the multivariate central limit theorem can be applied.

As an illustration, consider again the multivariate Gaussian distribution. As indicated in Krzanowski (1988, p. 226) and Christensen (1991, p. 26), if $\mathbf{X}$ follows a multivariate Gaussian distribution with unknown vector of expected values $\vec{\mu}$ and covariance matrix $\mathbf{\Sigma}$, and we use the vector of arithmetic means to estimate $\vec{\theta} = \vec{\mu}$, we get $\bar{\mathbf{X}} = 1/N \sum_{n=1}^{N} \mathbf{X}_n$ and $\bar{\mathbf{X}}$ follows a multivariate Gaussian distribution with expected value $\vec{\mu}$ and covariance matrix $\mathbf{\Sigma}/N$. The density function of $\bar{\mathbf{X}}$ depends not only on the vector of expected values, but also on the covariance matrix. Therefore, for a random sample, we use the fact that

$$Q = \frac{N(N-D)}{D(N-1)}(\bar{\mathbf{X}} - \vec{\mu})'\widehat{\mathbf{S}}^{-1}(\bar{\mathbf{X}} - \vec{\mu}),$$

where $\widehat{\mathbf{S}}$ is the matrix of experimental covariances and follows Fisher's $\mathcal{F}$ distribution with $(D, N-D)$ degrees of freedom. Note that for $D = 1$, Q follows Fisher's $\mathcal{F}$ distribution with $(1, N-1)$ degrees of freedom, which is equivalent to saying that $\sqrt{Q}$ follows Student's $\mathtt{t}$ distribution with $(N-1)$ degrees of freedom, the result obtained in the univariate case.

Again, we can readily obtain $h(\vec{\mu}) = \mathcal{F}^{\alpha}_{(D,N-D)}$ from the tables of Fisher's $\mathcal{F}$ distribution for any desired value of α given a sample size N. As in the univariate case, the function is a constant, resulting in

$$\mathrm{P}\left[\frac{N(N-D)}{D(N-1)}(\bar{\mathbf{X}} - \vec{\mu})'\widehat{\mathbf{S}}^{-1}(\bar{\mathbf{X}} - \vec{\mu}) \leq \mathcal{F}^{\alpha}_{(D,N-D)}\right] = 1 - \alpha$$

or, equivalently,

$$\mathrm{P}\left[(\bar{\mathbf{X}}-\vec{\mu})'\widehat{\mathbf{S}}^{-1}(\bar{\mathbf{X}}-\vec{\mu}) \le \frac{D(N-1)}{N(N-D)}\mathcal{F}^{\alpha}_{(D,N-D)}\right] = 1-\alpha. \tag{6.17}$$

The expression $(\bar{\mathbf{X}}-\vec{\mu})'\widehat{\mathbf{S}}^{-1}(\bar{\mathbf{X}}-\vec{\mu}) = [\text{constant}]$, as a function of the vector of expected values $\vec{\mu}$, defines the boundary of a hyperellipsoid centered at $\bar{\mathbf{X}}$ in the space of admissible and possible values of $\vec{\mu}$. Note that this result is derived under the assumptions of multivariate normality for $\mathbf{X}$ and independence of the sample.

If $\mathbf{X}$ does not follow a multivariate Gaussian distribution, the multivariate central limit theorem establishes that if N is large, then $Q = (\bar{\mathbf{X}}-\vec{\mu})\big/\sqrt{\widehat{\mathbf{S}}/N}$ is approximately multivariate normal with expected value 0 and covariance matrix the identity matrix. This result is again of crucial importance, as it guarantees that in most cases a confidence region analogous to that given in equation (6.17) can be obtained for the vector of expected values of a random vector if it is estimated by the vector of arithmetic means of the sample. To obtain it we proceed as follows: Because the density function f_Q of Q is standard multivariate normal, it is completely specified and $h(\vec{\mu})$ is a constant, that is,

$$h(\vec{\mu}) = f_Q(Q) = \frac{1}{(2\pi)^{p/2}}\exp\left\{-\tfrac{1}{2}Q'Q\right\},$$

where p stands for the number of components of the random vector $\mathbf{X}$. Equivalently, we can write

$$Q'Q = -2\ln\left\{(2\pi)^{p/2}h(\vec{\mu})\right\}.$$

But $Q'Q = N(\bar{\mathbf{X}}-\vec{\mu})'\,\widehat{\mathbf{S}}^{-1}(\bar{\mathbf{X}}-\vec{\mu})$ and thus we obtain the expression

$$(\bar{\mathbf{X}}-\vec{\mu})'\,\widehat{\mathbf{S}}^{-1}(\bar{\mathbf{X}}-\vec{\mu}) = -\tfrac{2}{N}\ln\left\{(2\pi)^{p/2}h(\vec{\mu})\right\}.$$

Recall again that to say *in most cases* does not mean *in all cases*: the requirement of a large N is quite arbitrary, the degree of convergence to the multivariate Gaussian distribution depending heavily on the multivariate density of the parent population, especially when it is skewed. According to Krzanowski (1988), a standard value of 25 is assumed to be reasonable, although faster convergence is obtained for reasonably symmetric parent densities, while for heavily skewed densities convergence requires a much larger number of samples. Compositional data is a particularly extreme case. The hyperellipsoid for a composition with D components will necessarily be defined in D-dimensional real space, while the corresponding simplex has, by definition, one less dimension. It is clear that most of the points in the confidence region given by the hyperellipsoid are impossible realizations of the vector of expected values, since they do not satisfy the constant-sum constraint. In fact, the probability of the set of points which do satisfy the constraint, as they lie on the simplex, is zero. This point has been illustrated in Figure 6.1.

If we assume now that our composition $\mathbf{Z}$ follows an additive logistic normal distribution, as stated in definition 2.15, then we know that $\text{alr}(\mathbf{Z}) = \mathbf{X}$ follows a multivariate normal distribution and we can apply the theory developed in previous paragraphs. Once we have the appropriate confidence region for $\mathbf{X}$, we can readily obtain the corresponding confidence region in the simplex for $\mathbf{Z}$, using the inverse transformation agl. This confidence region, with a specified confidence level, will be completely inbedded in the simplex, thus avoiding the problem of impossible values. Compare Figure 6.1 with Figure 6.2 for an illustration of these assertions.

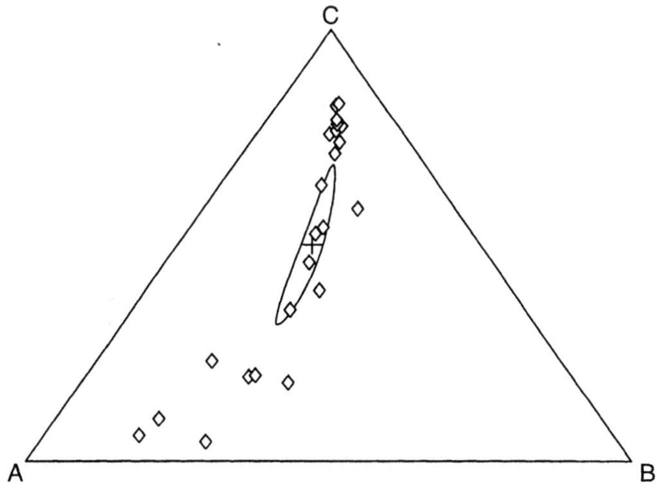

Figure 6.2: Confidence region in a ternary diagram.

As stated by Krzanowski (1988), the above theory concerns the standard approach to the one-sample estimation problem when applied to well-behaved data. Krzanowski mentions two situations that escape from it, namely that of missing values and of the presence of outliers. The interested reader is referred to Krzanowski (1988) for further details. We concentrate mainly on the case of dependent samples, which will be presented in the next section.

6.5.3 Confidence intervals and confidence regions in the regionalized case

The case of confidence intervals for regionalized variables in a standard approach has been studied by Dowd (1989) and by Tercan and Dowd (1995), while a model-based approach can be found in Diggle et al. (1998). The only publications known to the authors that are related to confidence regions for coregionalizations are by Cressie (1991) and by ver Hoef and Cressie (1993). Dowd (1989) presents two ways of viewing the problem of estimating confidence intervals: as probability intervals obtained from estimated conditional distributions; and, as the classical statistical method of inference from the sampling distribution of the estimator, following the line presented in Subsection 6.5.2. Tercan and Dowd (1995) present a short review of the state of the art related to probability intervals as obtained from an indicator, multi-Gaussian or disjunctive-kriging approach for the point support and extend the methodology to situations where some form of a change-of-support model is required. Diggle et al. (1998) give a theoretical approach under Gaussian assumptions both for linear estimators and generalized linear estimators in the univariate case; ver Hoef and Cressie (1993) mention the necessity of constructing a multidimensional prediction region around the predicted vector in the context of multivariable spatial prediction using a new cross-semivariogram.

Here we will not use the probability-interval approach, as it goes beyond the scope of the monograph, but instead we will consider an approximate solution to the inference from the sampling distribution of the estimator in an analogous approach to that given by Dowd (1989) and by Diggle et al. (1998), but concentrating on a presentation that follows the previous exposition of results for confidence intervals and regions in the nonregionalized case.

Consider the regionalized variable $X(\mathbf{u})$ and a sample $X(\mathbf{u}_1)$, $X(\mathbf{u}_2)$, ..., $X(\mathbf{u}_N)$ of size N. Note that in the following steps the sample will not be a random sample, i.e., the random variables $X(\mathbf{u}_1)$ will not be independent inasmuch as they are mutually correlated in most cases, the degree of correlation being measured by their spatial autocovariance function. We are interested in finding a confidence interval or interval estimate for $X(\mathbf{u}_0)$ at an unsampled location $\mathbf{u}_0$. According to the previous section, we need a function Q of the sample with known density such that either the function itself or the density depends on the parameter to be estimated, in this case the unknown value $X(\mathbf{u}_0)$. It is standard practice to take as an estimator $Q = X^*(\mathbf{u}_0)$ of the unknown value $X(\mathbf{u}_0)$ the one which minimizes the estimation mean-square error $\mathrm{E}[(X(\mathbf{u}_0) - X^*(\mathbf{u}_0))^2]$. The point estimator which accomplishes this condition for any arbitrary distribution of the parent process is known as the *kriging* *estimator*, which is

$$X^*(\mathbf{u}_0) = \mathrm{E}[X(\mathbf{u}_0)|X(\mathbf{u}_1), X(\mathbf{u}_2), ..., X(\mathbf{u}_N)],$$

with associated estimation variance

$$\sigma_0^2 = \mathrm{Var}[X(\mathbf{u}_0)|X(\mathbf{u}_1), X(\mathbf{u}_2), ..., X(\mathbf{u}_N)].$$

The next step in constructing a confidence interval for the predicted value requires forming an hypothesis on the distribution of the parent process. But as stated by Dowd (1989), this solution will in general lead to practically intractable expressions, and an approach based on multi-Gaussian transforms that incorporates bounds present for the variables under study is suggested. As a consequence, the solution lies in transforming the data distribution to the standard Gaussian distribution, estimating on the basis of transformed values, and then applying appropriate inverse transforms to the estimate and the confidence interval.

Let us assume that $X(\mathbf{u})$ is second-order stationary Gaussian. In this case $X^*(\mathbf{u}_0)$ is a linear function of the sample, leading to the well-known expression

$$X^*(\mathbf{u}_0) = \sum_{n=1}^{N} \lambda_n X(\mathbf{u}_n).$$

Furthermore, $X^*(\mathbf{u}_0)$ follows a normal distribution with expected value $\mu = \mathrm{E}[X(\mathbf{u})]$ and variance

$$\sigma_0^2 = \sum_{n=1}^{N}\sum_{\ell=1}^{N} \lambda_n \lambda_\ell \mathrm{Cov}[X(\mathbf{u}_n), X(\mathbf{u}_\ell)],$$

where σ_0^2 is a theoretical value that depends on the covariances $\mathrm{Cov}[X(\mathbf{u}_n), X(\mathbf{u}_\ell)]$, which under the hypothesis of stationarity depend only on the distances $\mathbf{h}_{n\ell} = |\mathbf{u}_n - \mathbf{u}_\ell|$. Thus, to follow the method presented in the nonregionalized case, we need an estimator of σ_0^2, a reasonable candidate being the kriging variance

$$\sigma_{\kappa 0}^2 = \sum_{n=1}^{N}\sum_{\ell=1}^{N} \lambda_n \lambda_\ell \mathrm{Cov}_m(X(\mathbf{u}_n), X(\mathbf{u}_\ell)),$$

where Cov_m stands for the model chosen to describe the spatial covariance structure of the parent population. Using this estimate, we can construct an approximate confidence interval for the unknown value assuming that the random variable $(X^*(\mathbf{u}_0) - X(\mathbf{u}_0))/\sigma_{\kappa 0}$ will reasonably approximate a Gaussian distribution under the stated hypothesis. Consequently, an approximate 95%

confidence interval will be obtained by using the realization of $X^*(\mathbf{u}_0)$ plus or minus $1.96 \cdot \sigma_{\kappa 0}$.

There are several aspects of this approximation of which we need to be aware. It is standard practice to use a Gaussian approximation, although for small samples we might be tempted to use Student's t. Although the main point would be the proof of the independence of the variance estimator used, $\sigma^2_{\kappa 0}$, and the kriging estimator, $X^*(\mathbf{u}_0)$, or finding explicitly the density of the random variable $(X^*(\mathbf{u}_0) - X(\mathbf{u}_0))/\sigma_{\kappa 0}$, there are several points that require proof in the process of arriving at a theoretically sound position. This is not an easy task, because $\sigma^2_{\kappa 0}$ depends on the sample through the estimators of the parameters of the spatial covariance structure model. This fact led Diggle et al. (1998) to assert that a potential risk in using the Gaussian confidence interval lies in the fact that it takes the estimation variance as an estimate of precision without taking into account the uncertainty involved in estimating the parameters of the assumed covariance structure.

There is also a tendency, based on the central limit theorem, to expect the distribution of $(X^*(\mathbf{u}_0) - X(\mathbf{u}_0))/\sigma_{\kappa 0}$ to be Gaussian. The flaw in this assertion lies in the fact that the central limit theorem requires independence of the sample, the negation of which is the basis of the theory of regionalized variables. In any case, even if the Gaussian assumption approximately holds, the usual number of samples involved in kriging is too small to expect a reasonable convergence to normality when the underlying distribution is skewed. This is easily illustrated by considering data with a skew histogram, which can be modeled reasonably by a lognormal density. In this case a confidence interval would be obtained as follows: Using the fact that a log-transform Y for a lognormal variable X follows a Gaussian distribution, let $Y^*(\mathbf{u}_0)$ be the simple kriging estimator with estimation variance $\sigma^2_{\kappa 0}$. An approximate 95% confidence interval for $(Y^*(\mathbf{u}_0) - Y(\mathbf{u}_0))/\sigma_{\kappa 0}$ would thus be obtained by substituting the realization of $Y^*(\mathbf{u}_0)$ into the expression

$$\left[Y^*(\mathbf{u}_0) - 1.96\sigma_{\kappa 0} \leq Y(\mathbf{u}_0) \leq Y^*(\mathbf{u}_0) + 1.96\sigma_{\kappa 0}\right],$$

inasmuch as the assumptions imply that

$$\begin{aligned} & \mathrm{P}\left[Y^*(\mathbf{u}_0) - 1.96\sigma_{\kappa 0} \leq Y(\mathbf{u}_0) \leq Y^*(\mathbf{u}_0) + 1.96\sigma_{\kappa 0}\right] \\ = \; & \mathrm{P}\left[-1.96 \leq \frac{Y^*(\mathbf{u}_0) - Y(\mathbf{u}_0)}{\sigma_{\kappa 0}} \leq +1.96\right] \\ = \; & 0.95. \end{aligned}$$

Analogous arguments hold for the multivariate case and lead to the conclusion that, to be on the safe side, it is always better to use transformed values whenever the sample histogram—or theoretical reasons, as in the case of constrained data—assure us that a hyperellipsoidal confidence region is not a reasonable solution, particularly when it leads to impossible or unreasonable results (e.g., negative values for positive random variables, or vectors that do not satisfy the constant-sum constraint in the case of compositional data).

6.6 H-data files

Consider for a given lag $\mathbf{h}$ the vector

$$\mathbf{Z}(\mathbf{h}) = \begin{pmatrix} \mathbf{Z}(\mathbf{u}) \\ \mathbf{Z}(\mathbf{u}+\mathbf{h}) \end{pmatrix} = \left(Z_1(\mathbf{u}) \ \cdots \ Z_D(\mathbf{u}) \ \ Z_1(\mathbf{u}+\mathbf{h}) \ \cdots \ Z_D(\mathbf{u}+\mathbf{h}) \right)' \tag{6.18}$$

and the vector of the alr transformed compositions at points $\mathbf{u}$ and $\mathbf{u}+\mathbf{h}$,

$$\mathbf{Y}(\mathbf{h}) = \begin{pmatrix} \mathbf{Y}(\mathbf{u}) \\ \mathbf{Y}(\mathbf{u}+\mathbf{h}) \end{pmatrix} = \begin{pmatrix} \operatorname{alr}(\mathbf{Z}(\mathbf{u})) \\ \operatorname{alr}(\mathbf{Z}(\mathbf{u}+\mathbf{h})) \end{pmatrix}. \tag{6.19}$$

The usual covariance matrix for $\mathbf{Y}(\mathbf{h})$ will have a four-block structure:

$$\operatorname{Cov}[\mathbf{Y}(\mathbf{h})] = \begin{pmatrix} \boldsymbol{\Sigma}(0,0) & \boldsymbol{\Sigma}(0,\mathbf{h}) \\ \boldsymbol{\Sigma}(\mathbf{h},0) & \boldsymbol{\Sigma}(\mathbf{h},\mathbf{h}) \end{pmatrix}, \tag{6.20}$$

where

$$\begin{aligned} \boldsymbol{\Sigma}(0,0) &= \operatorname{Cov}[\mathbf{Y}(\mathbf{u}), \mathbf{Y}(\mathbf{u})], \\ \boldsymbol{\Sigma}(0,\mathbf{h}) &= \operatorname{Cov}[\mathbf{Y}(\mathbf{u}), \mathbf{Y}(\mathbf{u}+\mathbf{h})], \\ \boldsymbol{\Sigma}(\mathbf{h},0) &= \operatorname{Cov}[\mathbf{Y}(\mathbf{u}+\mathbf{h}), \mathbf{Y}(\mathbf{u})], \\ \boldsymbol{\Sigma}(\mathbf{h},\mathbf{h}) &= \operatorname{Cov}[\mathbf{Y}(\mathbf{u}+\mathbf{h}), \mathbf{Y}(\mathbf{u}+\mathbf{h})]. \end{aligned}$$

Nonregionalized alr covariance matrices appear in the two blocks along the main diagonal. Both should be identical if no drift is present and the hypothesis of second-order stationarity cannot be rejected. Both blocks on the minor diagonal contain the alr cross-covariances, once for $\mathbf{h}$ and once for $-\mathbf{h}$. The covariance matrix $\operatorname{Cov}[\mathbf{Y}(\mathbf{h})]$ in (6.20) is, for every lag $\mathbf{h} \in H$, a symmetric matrix of variances and covariances. It differs from the usual covariance matrices through its special substructure. It is essential for spatial-structure analysis.

For numerical evaluation it is necessary to derive a new data set for every lag $\mathbf{h}$ and each direction of interest. This is performed in the following manner: For a given lag $\mathbf{h}$ all pairs of points in the initial data set that lie in the direction of $\mathbf{h}$ at a distance $\mathbf{h}$ are determined. For a given direction of $\mathbf{h}$, the starting points $\mathbf{Z}(\mathbf{u})$ and the end points $\mathbf{Z}(\mathbf{u}+\mathbf{h})$ are selected in the same manner as in the calculation of the semivariograms of residuals. From these data the vectors $\mathbf{Y}(\mathbf{u})$ are formed. The resulting sample contains twice the number of variables. In this way, however, conventional methods can be used for further analysis. The construction of the new data set should be based on the same principles used in calculation of semivariograms—tolerance in the lag and angular direction. The intervals $\mathbf{h}$ to be chosen depend on the sampling grid and the statement of the problem. In the following, the data files (called $\mathbf{h}$-*data files*, for short) are constructed in this manner.

6.7 Criteria for comparing results

Whenever different methods are applied, there is a need to define which criteria will be employed to compare the final results. In univariate geostatistics, predictive sample reuse based on the "leaving-one-out" technique known as *cross-validation* is traditionally used to find the best model for the semivariogram among several alternatives (Davis 1987). Cross-validation can also be used to assess the superiority of one method over another, as it allows us to measure which method gives estimates that are closer in some sense to observed samples (Olea 1999, chapter 7). This is our aim when defining appropriate distances or measures of discrepancy. Another possible approach consists of looking for a global measure of performance that can be found by analogy to multidimensional scaling. Both criteria are presented in the following.

6.7.1 Distance between observed samples and estimates

Cross-validation requires an appropriate discrepancy measure, yet there is no unique discrepancy measure known that would or could be universally applied. Thus, the desirable features for selection of such a measure should be ease of computation and interpretation. Ease of interpretation requires the measure of discrepancy (a) to conform to the sample space of the data set and (b) to be one-dimensional, whether the sample is uni- or multivariate. We are dealing with the latter case, as compositional data have several attributes; thus, we are interested in the combined goodness of the estimation of all of those attributes per sampling location.

Cross-validation must be used with caution, however. Complex interdependencies among the errors preclude finding the distribution for the errors, hindering a rigorous analysis or testing of results. In most situations geostatisticians employ the results as a guide in the analysis and estimation of attributes, however, special circumstances deserve extra caution:

1. The sampling pattern may lead to errors that are not representative of those that arise from an exhaustive sampling. A typical situation is that of sampling on clusters, where every sampling site will be close to at least one other sampling site. The cross-validation errors will be deceptively small, because for every site $\mathbf{u}_i$ there will always be a close site among the retained $n-1$ sites in the sampling. Cross-validation of such patterns will not include the large errors in the estimation for critical sites away from control points.

2. In the case of kriging, the estimate is insensitive to scaling of the semi-variogram, thus making cross-validation insensitive to changes in the sill (Olea 1999, pp. 24; 41).

Yet cross-validation remains a useful tool, not necessarily to prove anything but primarily to avoid blunders and draw honest conclusions by comparison.

Given a general coregionalization $\mathbf{Y}(\mathbf{u})$, whatever its sample space, the simplest approach consists in using an appropriate distance measure $\delta(\mathbf{Y}(\mathbf{u}), \mathbf{Y}^*(\mathbf{u}))$ between the observed vector $\mathbf{Y}(\mathbf{u})$ and its estimate $\mathbf{Y}^*(\mathbf{u})$ at each sampling location $\mathbf{u}$. The reason is that there are many well-known distance measures with known properties that are easy to compute, easy to interpret, and that conform to the sample space. They are also one-dimensional, inasmuch as a distance between two points in a given space is a single number, thus a multidimensional comparison in the case of several attributes is reduced to a unidimensional one. For any predictive method, the vector of distances between observed vectors and their estimates at each sampling point can be analyzed using standard descriptive statistics once an appropriate distance has been chosen. Furthermore, if we consider the vectors of distances obtained for two different estimation methods, say $\delta(\mathbf{Y}(\mathbf{u}), \mathbf{Y}_1^*(\mathbf{u}))$ and $\delta(\mathbf{Y}(\mathbf{u}), \mathbf{Y}_2^*(\mathbf{u}))$, their behavior can be compared using quantiles. It is also possible to consider the vector of differences between both,

$$\Delta(\mathbf{Y}_1^*, \mathbf{Y}_2^*) = \delta(\mathbf{Y}(\mathbf{u}), \mathbf{Y}_1^*(\mathbf{u})) - \delta(\mathbf{Y}(\mathbf{u}), \mathbf{Y}_2^*(\mathbf{u})). \tag{6.21}$$

If two estimation methods are very similar, $\Delta(\mathbf{Y}_1^*, \mathbf{Y}_2^*)$ should have a mean close to zero and the dispersion should not be too large. Conversely, for two very different estimation methods, the dispersion should be large. Also, the frequency distribution of $\Delta(\mathbf{Y}_1^*, \mathbf{Y}_2^*)$ and, in particular, the position of the mode can be very illustrative for verifying whether one estimation method performs better than the other. Note that although distances are by definition nonnegative numbers, $\Delta(\mathbf{Y}_1^*, \mathbf{Y}_2^*)$ can assume any value on the real line.

In the next chapter we will need different distances depending on whether we are working in real space or in the simplex. Let us consider first data in real space: Let $\mathbf{X}(\mathbf{u})$ be a coregionalization whose sample space is $\mathbb{R}^{D-1}$ and let $\mathbf{X}^*(\mathbf{u})$ be its estimate. Thus, the number of both observed and estimated components of $\mathbf{X}(\mathbf{u})$ is $D-1$ for each sampling point $\mathbf{u}$. A standard approach would be to use the Euclidean distance, whose square is defined as

$$\delta_e^2(\mathbf{X}(\mathbf{u}), \mathbf{X}^*(\mathbf{u})) = \sum_{i=1}^{D-1} (X_i(\mathbf{u}) - X_i^*(\mathbf{u}))^2 . \tag{6.22}$$

Nevertheless, Mahalanobis' distance is a more appropriate statistical distance in sample spaces where there exist differential variances and correlations between variates, as stated by Krzanowski (1988, p. 234). Moreover, this distance plays an important role in the case of compositional data, as we will discuss when distances on the simplex are introduced. For nonspatial statistics the squared Mahalanobis' distance is defined as

$$\delta_m^2(\mathbf{X}(\mathbf{u}), \mathbf{X}^*(\mathbf{u})) = (\mathbf{X}(\mathbf{u}) - \mathbf{X}^*(\mathbf{u}))\widehat{\mathbf{S}}^{-1}(\mathbf{X}(\mathbf{u}) - \mathbf{X}^*(\mathbf{u}))', \tag{6.23}$$

with the matrix $\widehat{\mathbf{S}}$ equal to the sample covariance matrix. In the spatial case, the question arises whether $\widehat{\mathbf{S}}$ should be replaced (1) by the sample covariance matrix obtained from the original data, (2) by the covariance matrix of the kriging or cokriging errors, or (3) by the covariance matrix estimated from the sills of the experimental semivariograms and cross-semivariograms. The appropriateness of the first option versus the last was discussed by Barnes (1991), who favors the latter if no trend appears in the data. The second option would be consistent with the usual approach in cross-validation, but it has more to do with the spatial dispersion of the samples used for estimation than estimating the variability of the components. In order to place more importance on the actual variability of the attributes, our choice is therefore the third option, which implies using the sill of the semivariograms as the proper estimate of the variances in the case of ordinary kriging, leading to a diagonal matrix, and replacing the off-diagonal zeros by the sill of the cross-semivariograms for ordinary cokriging. But, because of the fact that we are using the estimation procedure based on the Fast Fourier Transform (Yao and Journel 1998; Ma and Yao 2001), and because we have no explicit estimation of the sills, this option is not feasible. Thus, in practice, we have used as an approximation the variance–covariance matrix either of observations or of estimates as necessary.

On the simplex, however, it makes no sense to use either the Euclidean or the Mahalanobis' distance because they are translation invariant and do not satisfy basic requirements such as perturbation invariance and subcompositional dominance (Martín-Fernández et al. 1998b). Therefore, we will use the Aitchison distance δ_a and a compositional Mahalanobis' distance δ_c on the simplex. For a D-part regionalized composition, the Aitchison distance δ_a between the observed vector $\mathbf{Z}(\mathbf{u})$ and its estimate $\mathbf{Z}^*(\mathbf{u})$ is the square root of

$$\delta_a^2(\mathbf{Z}(\mathbf{u}), \mathbf{Z}^*(\mathbf{u})) = \sum_{i=1}^{D} \left(\ln \frac{Z_i(\mathbf{u})}{g(\mathbf{Z}(\mathbf{u}))} - \ln \frac{Z_i^*(\mathbf{u})}{g(\mathbf{Z}^*(\mathbf{u}))} \right)^2, \qquad (6.24)$$

where $g(\mathbf{Z}(\mathbf{u}))$ is the geometric mean of the components of $\mathbf{Z}(\mathbf{u})$ and $g(\mathbf{Z}^*(\mathbf{u}))$ of $\mathbf{Z}^*(\mathbf{u})$, that is,

$$g(\mathbf{Z}(\mathbf{u})) = \left(\prod_{i=1}^{D} Z_i(\mathbf{u}) \right)^{1/D} \quad \text{and} \quad g(\mathbf{Z}^*(\mathbf{u})) = \left(\prod_{i=1}^{D} Z_i^*(\mathbf{u}) \right)^{1/D}.$$

The squared compositional Mahalanobis' distance δ_c^2 is given by

$$\delta_c^2(\mathbf{Z}(\mathbf{u}), \mathbf{Z}^*(\mathbf{u})) = (\mathbf{X}(\mathbf{u}) - \mathbf{X}^*(\mathbf{u}))\, \widehat{\mathbf{S}}_{\mathbf{X}}^{-1} (\mathbf{X}(\mathbf{u}) - \mathbf{X}^*(\mathbf{u}))', \qquad (6.25)$$

where $\mathbf{X}(\mathbf{u}) = \text{alr}(\mathbf{Z}(\mathbf{u}))$ and $\mathbf{X}^*(\mathbf{u}) = \text{alr}(\mathbf{Z}^*(\mathbf{u}))$. Note that the matrix $\widehat{\mathbf{S}}_{\mathbf{X}}$ in (6.25) plays the same role as $\widehat{\mathbf{S}}$ in (6.23) and that the distance δ_c is identical to the clr Mahalanobis' distance mentioned in Martín-Fernández et al. (1998b), where its properties have been extensively analyzed.

6.7.2 STRESS between observed and estimated data set

A global measure of performance for comparing results is based on the concept of STRESS (STandardized REsidual Sum of Squares), which is used in multidimensional scaling (Jobson, 1992, p. 590; Cox and Cox, 1994, p. 45). There, STRESS is defined by the following equation:

$$\text{STRESS} = \left(\frac{\sum_{i<j} (\delta_{ij} - \delta_{ij}^*)^2}{\sum_{i<j} (\delta_{ij})^2} \right)^{1/2} \qquad (6.26)$$

where, for a general coregionalization $\mathbf{Y}(\mathbf{u})$, δ_{ij} is the distance between observations $\mathbf{Y}(\mathbf{u}_i)$ and $\mathbf{Y}(\mathbf{u}_j)$, and δ_{ij}^* is the distance corresponding to the same observations, but projected into a lower dimensional space as $\mathbf{Y}^*(\mathbf{u}_i)$ and $\mathbf{Y}^*(\mathbf{u}_j)$. The distances δ and δ^* are assumed to be of the same type, e.g., Mahalanobis' or Euclidean distance if the original space and the space of the projections are both real space, although with different dimensions. This index is zero when all δ_{ij}^* are equal to δ_{ij} and is one when all δ_{ij}^* are zero and at least one δ_{ij} is larger than zero. Thus, the smaller the STRESS, the more similar is the distance between original observations to the distance between projected observations, and therefore the loss of information due to the projection into the lower dimensional space is smaller. Here, we will use the same definition for STRESS, but δ^*_{ij} will be the distance between the corresponding estimates obtained using either ordinary kriging or ordinary cokriging.

To decide the appropriate distance to be used, recall that the concept of STRESS is an overall measure that compares the dispersion of the original data set with the dispersion of an estimated data set. It therefore seems reasonable to avoid over-weighting of attributes with a large variability, because this leads necessarily to the consideration of Mahalanobis' distance as the appropriate choice. Thus, for $\mathbf{Y}(\mathbf{u}) = \mathbf{X}(\mathbf{u}) \in \mathbb{R}^{D-1}$ we would have

$$\delta_{ij} = \delta_m(\mathbf{X}(\mathbf{u}_i), \mathbf{X}(\mathbf{u}_j)) \quad \text{and} \quad \delta^*_{ij} = \delta_m(\mathbf{X}^*(\mathbf{u}_i), \mathbf{X}^*(\mathbf{u}_j)), \tag{6.27}$$

and for $\mathbf{Y}(\mathbf{u}) = \mathbf{Z}(\mathbf{u}) \in \mathcal{S}^D$

$$\delta_{ij} = \delta_c(\mathbf{Z}(\mathbf{u}_i), \mathbf{Z}(\mathbf{u}_j)) \quad \text{and} \quad \delta^*_{ij} = \delta_c(\mathbf{Z}^*(\mathbf{u}_i), \mathbf{Z}^*(\mathbf{u}_j)). \tag{6.28}$$

It is important to note that, although $\widehat{\mathbf{S}}$ in (6.23) and $\widehat{\mathbf{S}}_{\mathbf{X}}$ in (6.25) are the same as the matrices required in computation of equations (6.27) and (6.28), Mahalanobis' distances used in computing the STRESS are not the same as the distances used for cross-validation in the previous approach. The distances used for computing STRESS are distances between vectors of the same data set, either of observed or of estimated vectors, while the distances used in the first criterion are computed between vectors of different data sets, involving always two vectors—one observed and one estimated—at each sampling location. Therefore, conclusions based on one criterion need not necessarily be the same as conclusions based on the other.

Chapter 7

Application to real data

Methods for spatial correlation analysis and estimation of r-compositions introduced in the foregoing chapters are illustrated here by an example that draws upon real data taken from the Lyons West oil field located in west-central Kansas, USA. Data consist of core analyses of water saturation, saturated thickness and average reservoir porosity over the connate saturated interval at different locations in the Lyons West field. These data are used to compare different possible methods for predicting regionalized compositions. The methods we consider are:

1. a direct approach for estimating compositional variables derived from the original measurements;
2. the basis method, applicable only when there is a random function that can be regarded as the size or accumulation of the regionalized variable under study;
3. the logratio approach, using the additive logratio (*alr*) transformation.

Kriging and cokriging estimation methods will be considered for original compositions and for transformed data. Software used for statistical analyses include GSLIB, programs written by Ma and Yao (2001) and *ad hoc* programs written by the authors. GSLIB is a public-domain library of geostatistical programs written in Fortran (Deutsch and Journel 1998); the other programs are available from their authors.

7.1 The Lyons West oil field of Kansas

The Lyons West oil field is located at 98° 15' west longitude and 38° 20' north latitude in west-central Kansas, near the center of the United States (Figure 7.1). The reservoir occurs in Mississippian (Lower Carboniferous) rocks

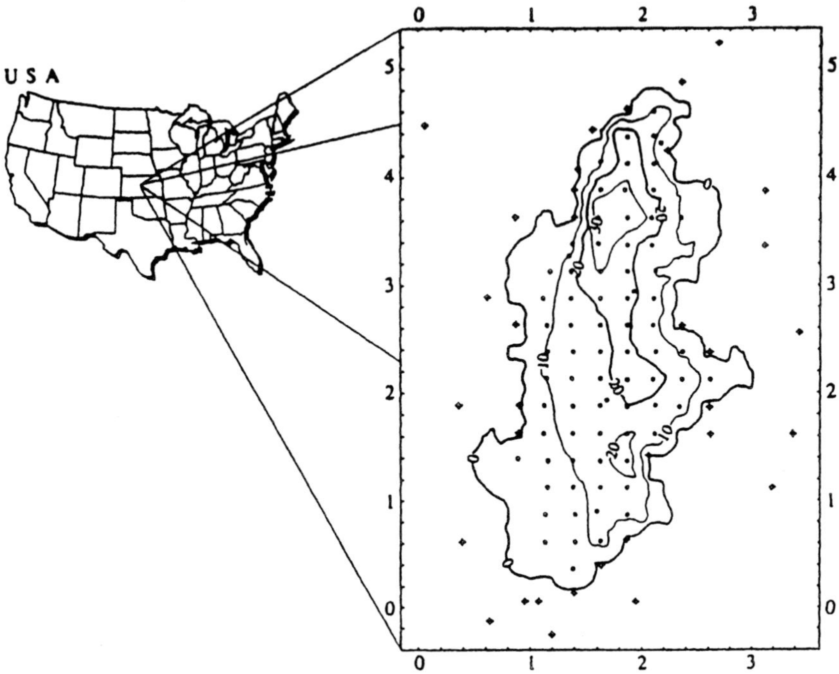

Figure 7.1: Wells in the Lyons West study area. Solid dots indicate location of producing wells used in the study; contour lines denote saturated thickness, in feet; x and y coordinates are in miles.

that originated as sediments deposited in the shallow interior sea that covered much of North America in the late Paleozoic. The field was discovered somewhat accidentally in 1963, during the drilling of a deeper Ordovician prospect. Initial oil in place was estimated at 22 million stock-tank barrels of oil.

7.1.1 Description of the field

The genesis of the reservoir, composed of carbonate-cemented sands, is interpreted as an offshore bar enclosed in marine shales. Regional uplift tilted the sand body, which was truncated along the western margins by the unconformity marking the base of the Pennsylvanian (Upper Carboniferous). The sandstones interfinger with marine shales to the east, but the eastern margin of the reservoir is defined by the intersection of the oil-water contact with the shale seal at the top of the reservoir interval (Ehm 1965).

7.1.2 Preparation of the oil-field data

The Lyons West oil field was developed on a fairly regular spacing and scheduled for water flooding. Data used here are part of a 1965 engineering study prepared for Kewanee Oil Company. While accurate measurements of porosity are easily achieved, drilling fluid ordinarily disturbs water-saturation values. Values used here were revised to conform to Buckles' equation for bulk volume of water (Buckles 1965), which in the case of the Lyons West field turned out to follow an equation of type III,

$$\varphi S_w = 181.9 - 6.06 S_w \, , \tag{7.1}$$

illustrated in Figure 7.2. Porosity, in percentage, is denoted by φ; S_w is connate water saturation in percentage. The revised values for water saturation (Figure 7.2) include perturbation by an amount inversely proportional to saturated thickness.

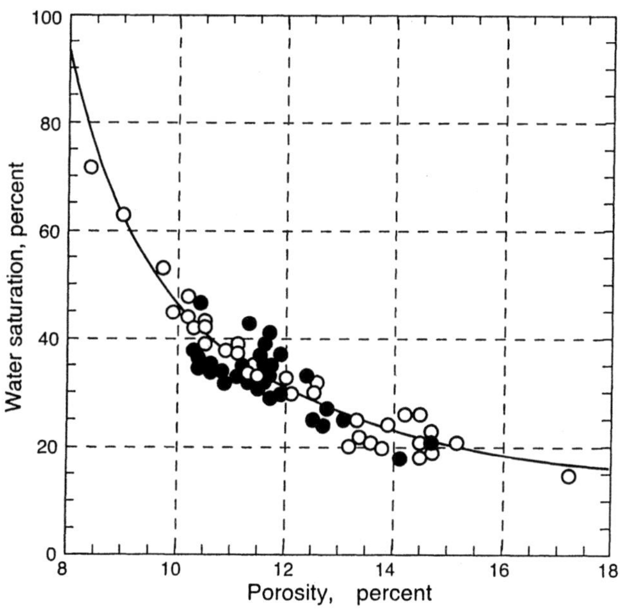

Figure 7.2: Revised Lyons West data. Solid dots denote original data used to define the parameters of Buckles' equation along the segmented line. Open circles denote revised data.

We can employ values of connate water saturation and porosity to define a compositional data set comprising the proportion of substances making up the

Lyons West reservoir: $\mathbf{Z}(\mathbf{u}) = (\text{oil}, Z_o(\mathbf{u}); \text{water}, Z_w(\mathbf{u}); \text{and rock}, Z_r(\mathbf{u}))$. In the case of Lyons West, the attributes of interest are:

$$\begin{aligned} Z_o(\mathbf{u}) &= 0.01\varphi(\mathbf{u})(100 - S_w(\mathbf{u})), \\ Z_w(\mathbf{u}) &= 0.01\varphi(\mathbf{u})S_w(\mathbf{u}), \\ Z_r(\mathbf{u}) &= 100 - \varphi(\mathbf{u}). \end{aligned} \tag{7.2}$$

The composition is a ternary system, thus it can be represented by ternary diagrams, as shown in Figure 7.3. The usual univariate descriptive statistics of the compositional variables are summarized in Table 7.1.

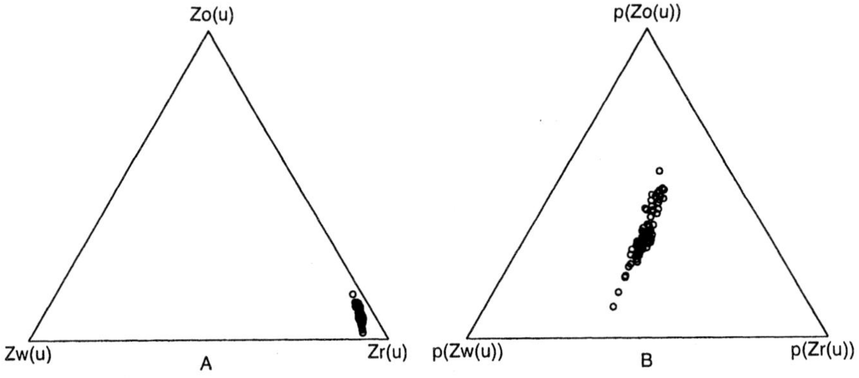

Figure 7.3: Representation of Lyons West compositional data in ternary diagrams. (A) Original data. (B) Centered data.

Note that because of the high proportion of component rock, original data plot in the lower right corner of diagram (A) in Figure 7.3. For a better visualization, we have centered the data set in diagram (B) following the procedure described in Martín-Fernández et al. (1999) and Buccianti et al. (1999), using the perturbation vector $\mathbf{p} = (0.32, 0.65, 0.03)'$. As a result, the baricenter of the original data set is moved to the baricenter of the triangle; in this way, relationships between variables can be better understood without modifying their statistical properties by using a logratio approach, inasmuch as perturbation in the simplex is equivalent to translation in real space (Aitchison 1997). Note that the quasi-constant behavior of $Z_r(\mathbf{u})$ is now evident, and also that small variations in either of the other two components, $Z_o(\mathbf{u})$ and $Z_w(\mathbf{u})$, cause the data to move from one extreme of the cloud to the other. This has the effect that the Aitchison distance and the compositional Mahalanobis' distance defined in (6.24) and (6.25) are sensitive to small variations in the estimation

Table 7.1: Univariate descriptive statistics for original composition.

Attribute	$Z_o(\mathbf{u})$	$Z_w(\mathbf{u})$	$Z_r(\mathbf{u})$
Minimum	2.26	2.52	82.78
1st quartile (Q_1)	6.73	3.50	87.35
Median (Q_2)	7.50	3.78	88.50
Mean	7.96	3.87	88.18
3rd quartile (Q_3)	9.35	4.30	89.33
Maximum	14.62	6.02	91.73
Std. deviation	2.18	0.70	1.60
$(Q_3 - Q_1)/2$	1.31	0.40	0.99
Skewness	0.45	0.43	-0.82

of these components, which is consistent with the compositional character of the data.

7.2 Direct estimation

As discussed, one of the essential characteristics of regionalized compositions is that their sample space is the simplex. In this section, using the Lyons West data set, we illustrate the problems arising from this fact and the effects it has when direct estimation is applied. First, considering the multivariate nature of any composition, the estimation method of choice would seem to be ordinary cokriging. However, the straightforward use of ordinary cokriging is not acceptable because the constant-sum condition implies that there is a linear dependency among the variables which will produce singular covariance matrices.

Secondly, if one resorts to estimation of each attribute separately using some form of kriging, there are still problems because kriging does not consider boundary constraints in determining the weights. Therefore it is possible for estimates to fall outside the range of the data. Conceivably, estimates may be negative or exceed the value of the sum—this did not happen in our study, largely because of the smoothing characteristic of kriging (Olea 1999, p. 141).

The third consideration is that direct application of kriging cannot constrain estimates of compositional attributes so they sum to a constant, violating an essential characteristic of compositional data. The situation is even worse for simulated realizations, which always have wider dispersion than kriging estimates. The sum of the three components from leaving-one-out estimates used for cross-validation of the model of the 76 sites considered for the Lyons West regionalized composition are shown in Figure 7.4. The sum of each of the three components for one partial, conditional realization comprising a grid of 11 columns and 37 rows obtained through simulated annealing is also shown in this illustration.

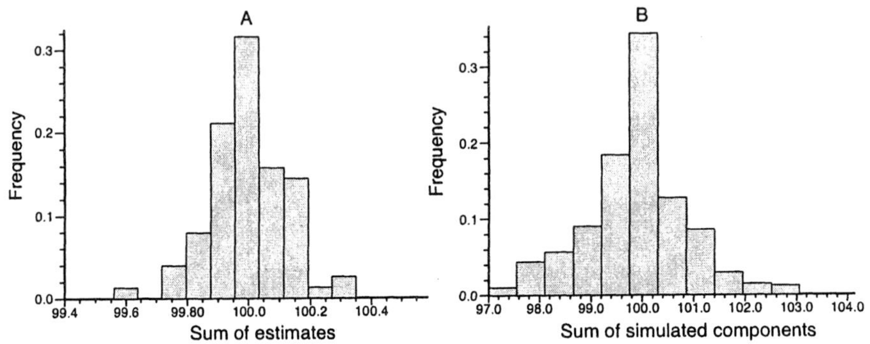

Figure 7.4: Histogram for the sum of three components of the compositional data, Lyons West field: (A) Cross-validation estimates. (B) Realization for the coregionalization using simulated annealing.

"Quick fixes" are unsatisfactory and lack theoretical justification. For example, if one compositional attribute is ignored in the estimation process, and then estimates for the censored attribute are made equal to the constant minus the sum of the other estimated values, the spurious nature of the cross-covariance used in the ordinary cokriging equations has not been eliminated. No matter what estimation method is used, results will differ depending upon which attribute has been removed from the estimation. Moreover, estimates for the censored attribute may still assume negative values if the sum of the estimates of the other attributes is larger than the value of the constant sum. Note that these arguments hold also for a coregionalization of proportions that do not sum to a constant, as a "fill-up" component can always be defined.

All of these problems associated with direct analysis, estimation and simulation of regionalized compositional data preclude serious consideration of the direct approach and call for alternative methods, like those described in previous chapters and used in Pawlowsky et al. (1995) to overcome their drawbacks.

Instead of using a transformation, one conceivably could introduce the necessary constraints into the ordinary cokriging system of equations to accommodate the peculiarities of compositional data. De Gruijter et al. (1997)

proposed such a generalization—called *compositional kriging* and based on ordinary kriging—which they admit is less than optimal, slow to run and cumbersome. The existence of any similar approach based on ordinary cokriging is unknown to the authors. The singularity of the system of equations could be overcome using generalized inverses. But to our understanding, the essential difficulty, which is the drawback in the case of compositional kriging, lies in the interpretation of error variances. Individual error variances are understood as confidence intervals around the estimated value of the corresponding component that could easily exceed acceptable limits for compositional data. Moreover, the joint error variances form a multidimensional parallelepiped that is not embedded in the simplex, thus making no sense at all. For all these reasons, judicious use of transformations remains as the best alternative.

7.3 The alr method

Aitchison (1986) has been a strong advocate of the additive logratio (alr) transformation introduced in Chapter 2 (definition 2.10) as a convenient way to analyze compositional data in nonspatial statistics. We will use this approach to analyze real data from our Lyons West study area—treating the data as regionalized compositions.

7.3.1 The particular case of Lyons West

As argued in Section 2.7 and Chapter 3 and developed in Chapter 5 (particularly in Section 5.4), the alr approach is also justified for regionalized compositions. In our case, the alr transformed variables of choice $X_1(\mathbf{u})$ and $X_2(\mathbf{u})$ are:

$$X_1(\mathbf{u}) = \ln \frac{Z_o(\mathbf{u})}{Z_r(\mathbf{u})}, \qquad X_2(\mathbf{u}) = \ln \frac{Z_w(\mathbf{u})}{Z_r(\mathbf{u})}. \tag{7.3}$$

One may notice that there is, as should be, one less transformed variable than compositional attributes. An important property of the transformation is the independence of results from the selection of the denominator in the alr transformation, a property known as *permutation invariance* and proved in Section 5.4.4. In this instance its inverse, the additive generalized logistic transformation (agl), for estimates is:

$$\begin{aligned} Z_o^*(\mathbf{u}) &= \frac{100\exp\left(X_1^*(\mathbf{u})\right)}{1+\exp\left(X_1^*(\mathbf{u})\right)+\exp\left(X_2^*(\mathbf{u})\right)}, \\ Z_w^*(\mathbf{u}) &= \frac{100\exp(X_2^*(\mathbf{u}))}{1+\exp\left(X_1^*(\mathbf{u})\right)+\exp\left(X_2^*(\mathbf{u})\right)}, \\ Z_r^*(\mathbf{u}) &= \frac{100}{1+\exp\left(X_1^*(\mathbf{u})\right)+\exp\left(X_2^*(\mathbf{u})\right)}. \end{aligned} \tag{7.4}$$

As mentioned in Section 2.7, the logarithmic transformation usually helps approximate normality, the condition at which kriging methods work best in a minimum mean-square-error sense (Diggle et al. 1998). As shown in Table 7.2, the Lyons West data follow the rule on average. For further details on the method, see Olea (1999).

Table 7.2: Maximum deviation between the experimental cumulative distribution function of an attribute and the cumulative distribution function of a normal variate with the same mean and variance. All samples contain 76 observations.

Attribute	$Z_o(\mathbf{u})$	$Z_w(\mathbf{u})$	$Z_r(\mathbf{u})$	$X_1(\mathbf{u})$	$X_2(\mathbf{u})$
Deviation	0.149	0.074	0.174	0.102	0.078

7.3.2 Spatial correlation

We have conducted the numerical modeling of covariances following the method and procedure explained in Section 6.2. In this instance, there are two variables, $X_1(\mathbf{u})$ and $X_2(\mathbf{u})$, hence there are two correlograms and one cross-correlogram. Figure 7.5 displays the correlogram maps that result with no overlapping of groups. Figure 7.6, presented in parts (A), (B), and (C), is a companion set of traverses to relate the maps to the more familiar directional graphical representations.

We employed program `corrmap` (Ma and Yao 2001) to generate grids for the maps in Figure 7.5, using the following parameters:

- Nodes along easting direction: 33 (*nxlag*=16)
- Nodes along northing direction: 33 (*nylag*=16)
- Internodal distance: 0.0667 mi
- Tolerance (*dxlag, dylag*): 0
- Minimum number of pairs: 3

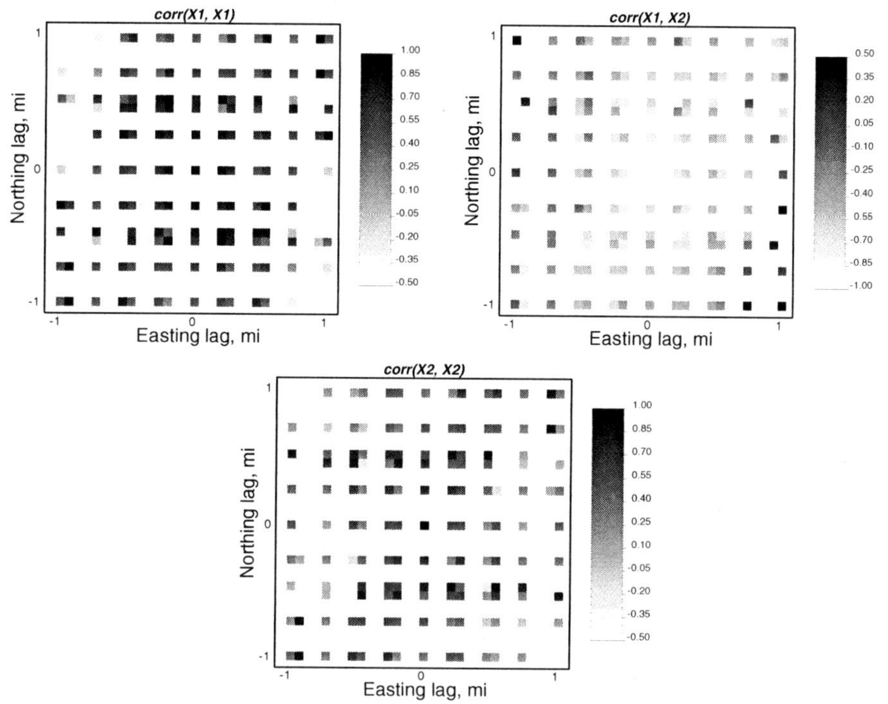

Figure 7.5: Correlogram maps for alr transformed data with no overlapping of groups.

Program **pixelplt** from GSLIB (Deutsch and Journel 1998) was used to plot the maps in Figure 7.5 and the following maps. Our own utility program was used to prepare data for the correlograms in Figure 7.6 (A–C) and following correlograms. The display itself was produced with a modified version of program **vargplt** in GSLIB. For comparison, Figure 7.6 (A–C) also displays the conventional directional correlogram calculated with program **gamv** in GSLIB with the following parameters:

- Angular tolerance (*atol*): 10°

- Band width (*bandw*): 0.5 mi

The unavoidable discrepancy between the values derived from the correlogram map generated with **corrmap** and the directional correlogram prepared with **gamv** is due to the way the different programs handle tolerance for the purpose of determining the distance classes. While **gamv** employs a polar search (Deutsch and Journel 1998, p. 49), program **corrmap** is based on a "pigeonhole" search (Ma and Yao 2001). Appropriate selection of parameters can minimize the discrepancies, as was done in Figure 7.6 (A–C). Figures 7.5 and 7.6 were generated primarily for control reference. Values in the grid are erratic because of the low number of pairs involved in each estimate and there are several values missing because the number of pairs ranged from 0 to 2.

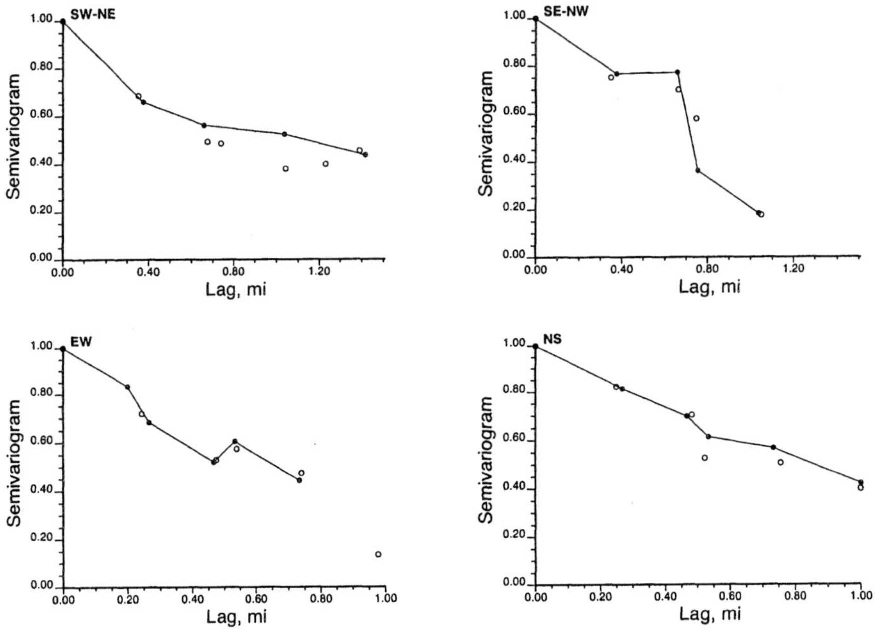

Figure 7.6: (A) Directional correlograms for X_1. Dots = values from maps in Fig. 7.5; open circles = conventional, experimental directional correlograms.

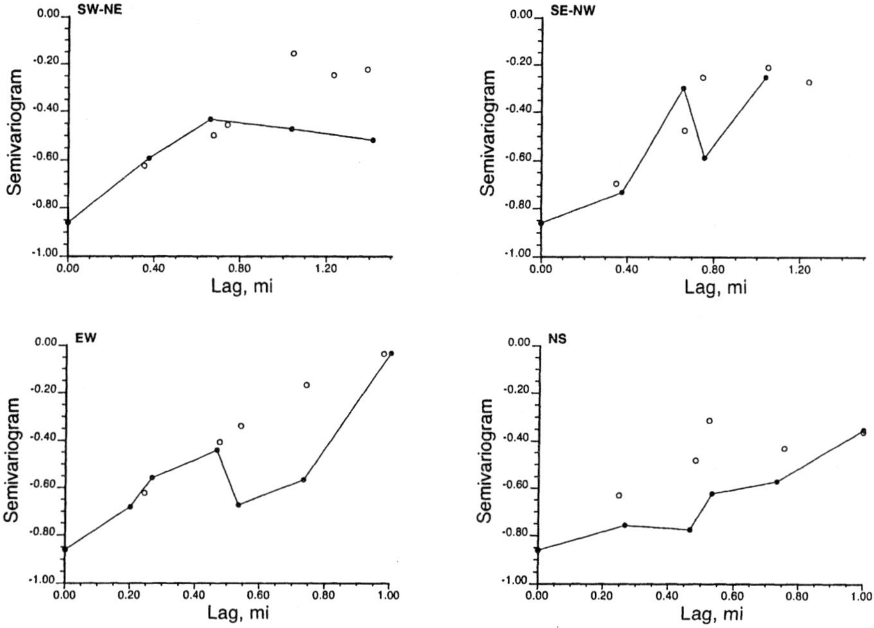

Figure 7.6: (B) Cross-correlograms between X_1 and X_2.

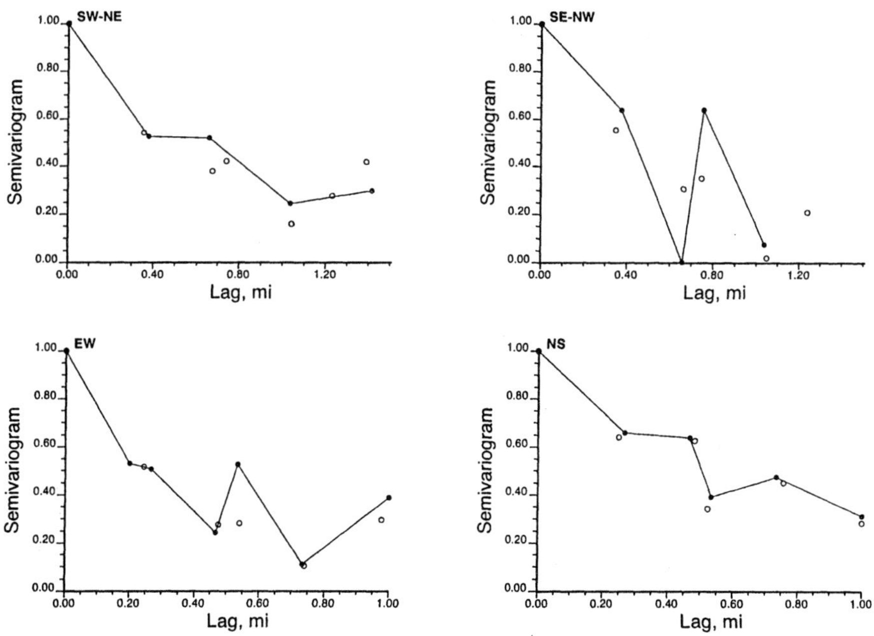

Figure 7.6: (C) Correlograms for X_2.

Figures 7.7 and 7.8 (A, B, C) are the true starting point in the modeling of permissible correlograms. The grids behind Figure 7.7 were prepared with the same program, `corrmap`, and the same parameters, with the exception of:

- Tolerance $(dxlag, dylag)$: 2
- Minimum number of pairs: 3

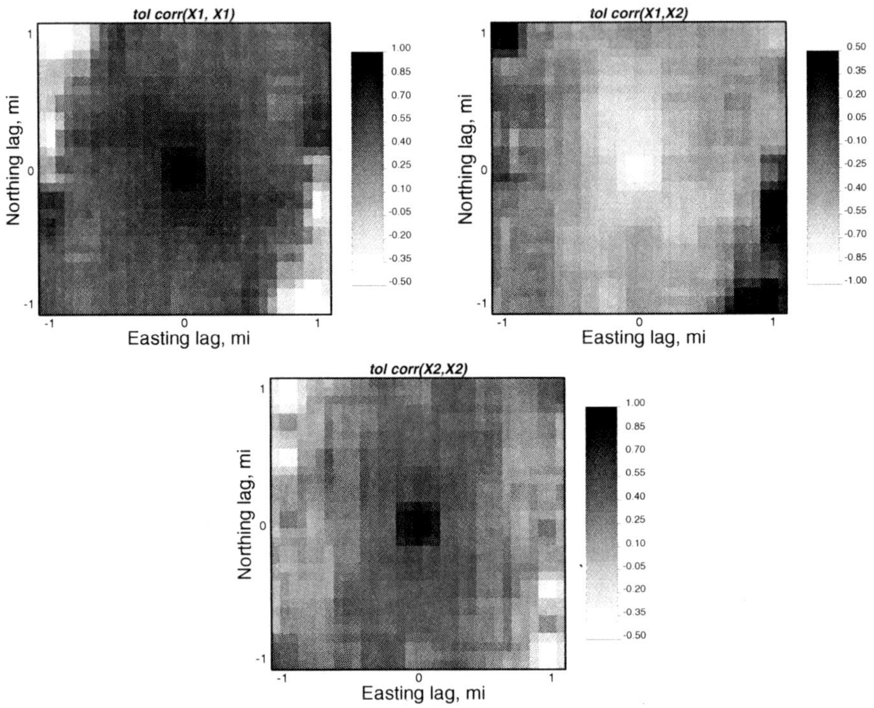

Figure 7.7: Correlogram maps for the alr transformation with use of overlapping.

The new averaging over 5-by-5 pigeonholes has the double effect of smoothing the maps and reducing the number of nodes without pairs below the threshold. Note that the practice of overlapping squares that contribute pairs to the estimates, along with the already different radial versus Cartesian search when $dxlag = dylag = 0$, deviates even further from the traditional estimation of experimental semivariograms, which is done over non-overlapping tessellations.

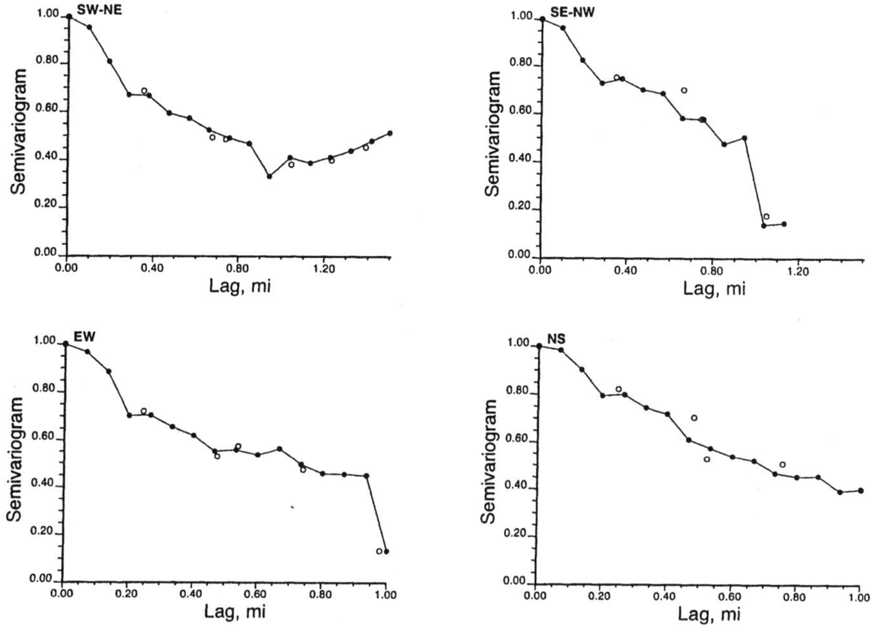

Figure 7.8: (A) Directional correlograms for X_1. Dots = values from maps in Fig. 7.7; open circles = conventional correlograms in Fig. 7.6 (A).

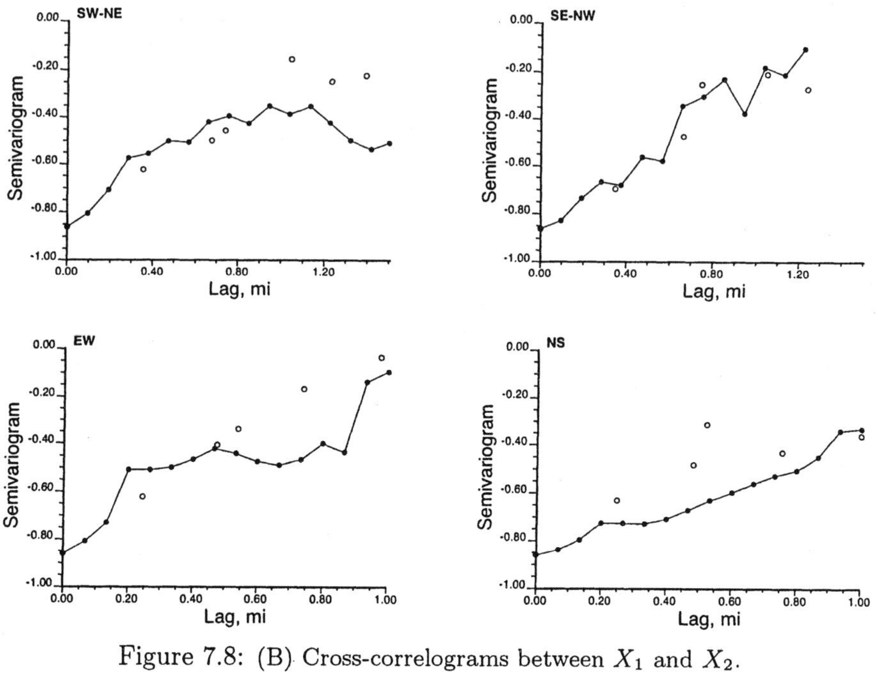

Figure 7.8: (B) Cross-correlograms between X_1 and X_2.

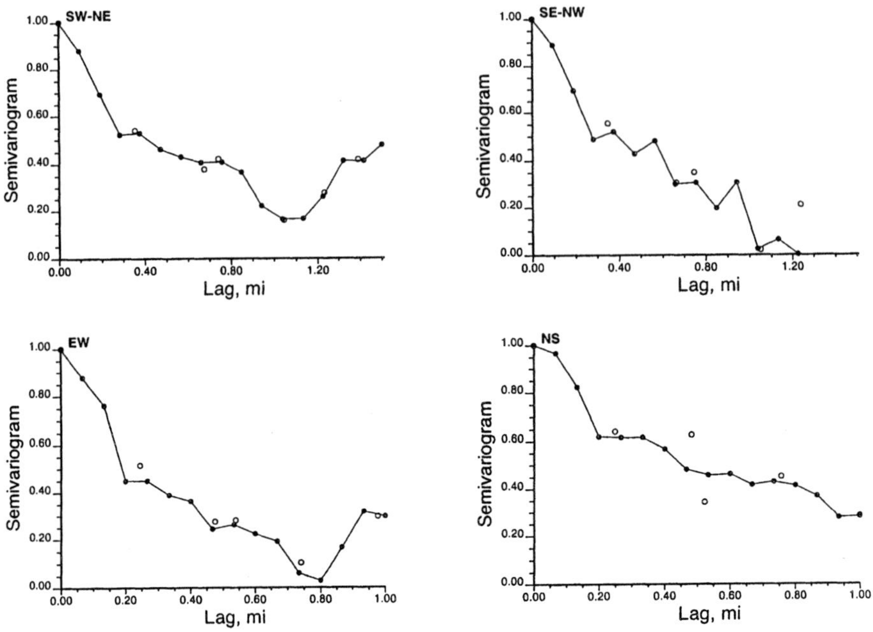

Figure 7.8: (C) Correlograms for X_2.

The next step in the production of permissible correlogram maps using the programs prepared by Ma and Yao (2001) is to run program `intmap`. This program interpolates any remaining missing values and also performs a preliminary smoothing. The parameters that we selected for the production of Figures 7.9 and 7.10 (A, B, C) were:

- Nodes along easting direction: 33
- Nodes along northing direction: 33
- Smoothing window size: 10
- Ratio of radius of fan: 0.1
- Gradual increment: 0.01
- Minimum number of data for smoothing: 5

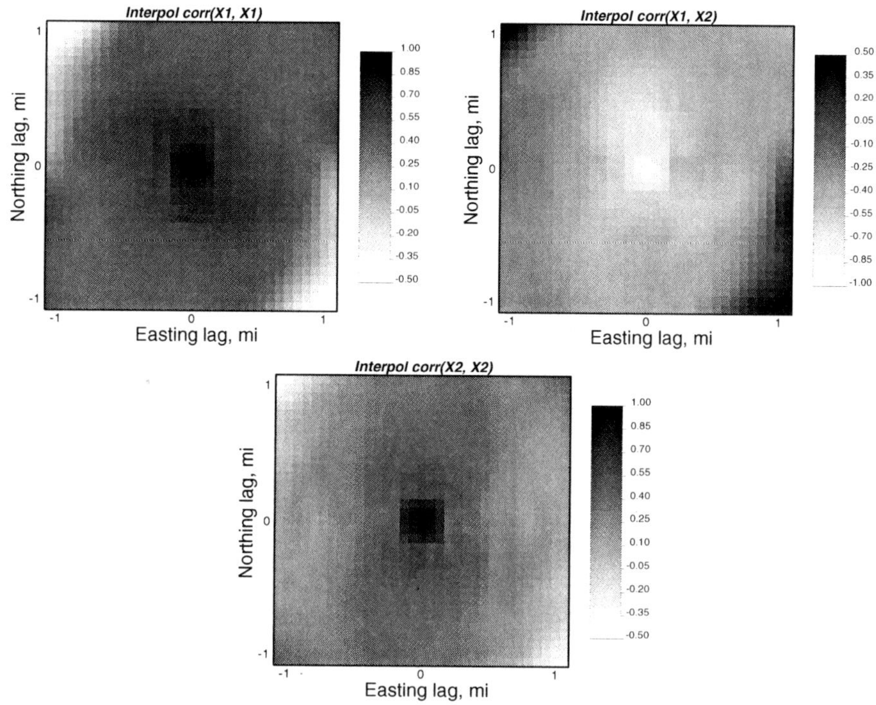

Figure 7.9: Correlogram maps for the alr transformation without missing values.

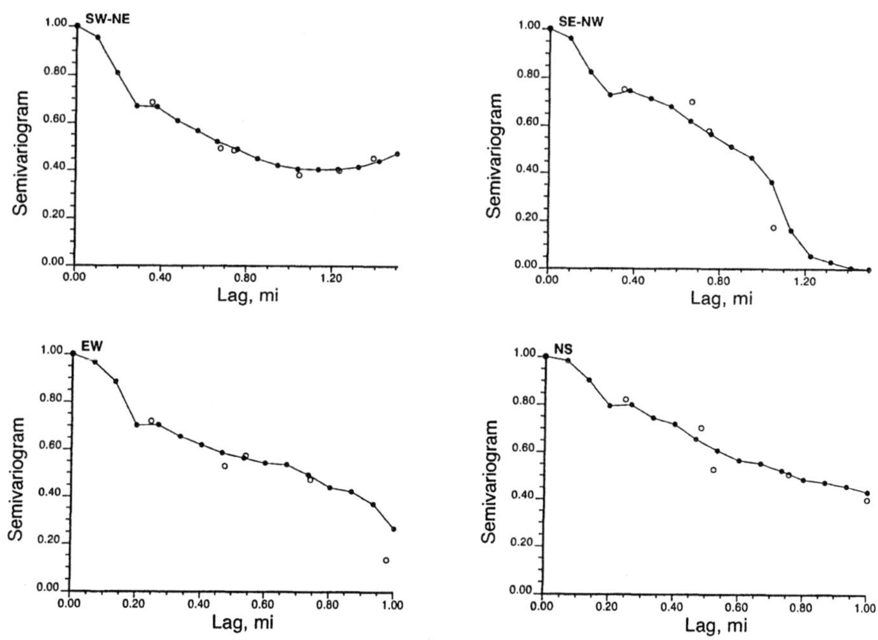

Figure 7.10: (A) Directional correlograms for X_1. Dots = values from maps in Fig. 7.9; open circles = conventional correlograms in Fig. 7.6 (A).

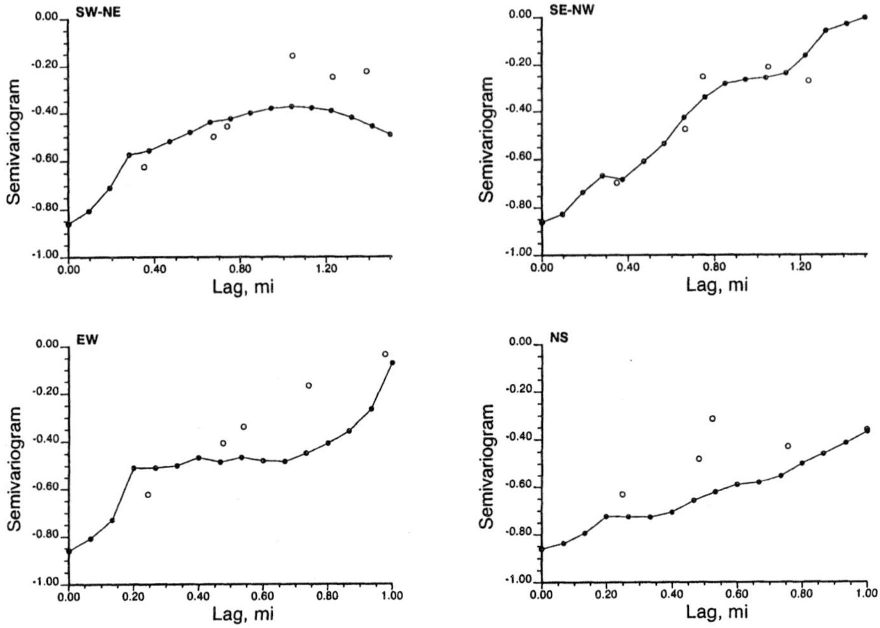

Figure 7.10: (B) Cross-correlograms between $X_1(\mathbf{u})$ and $X_2(\mathbf{u})$.

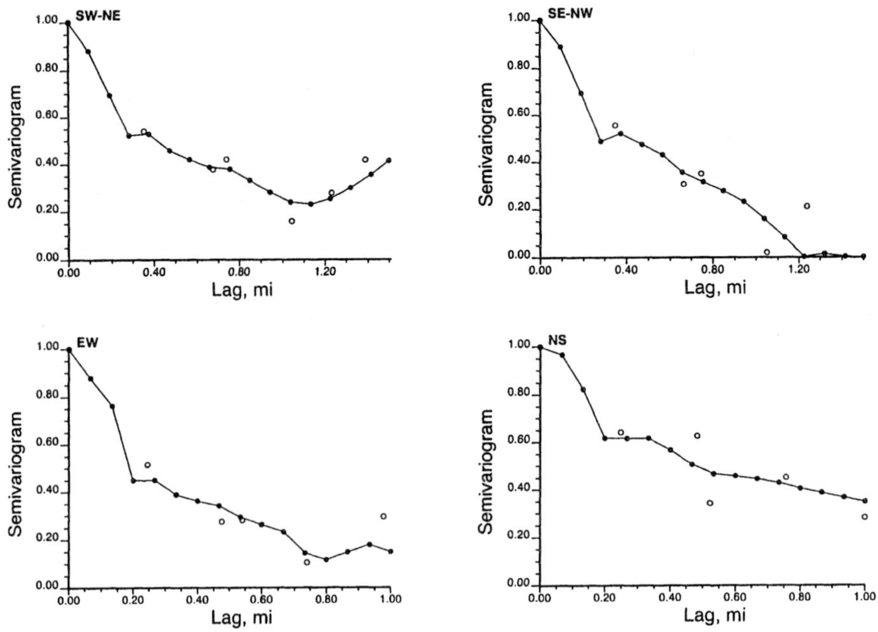

Figure 7.10: (C) Correlograms for $X_2(\mathbf{u})$.

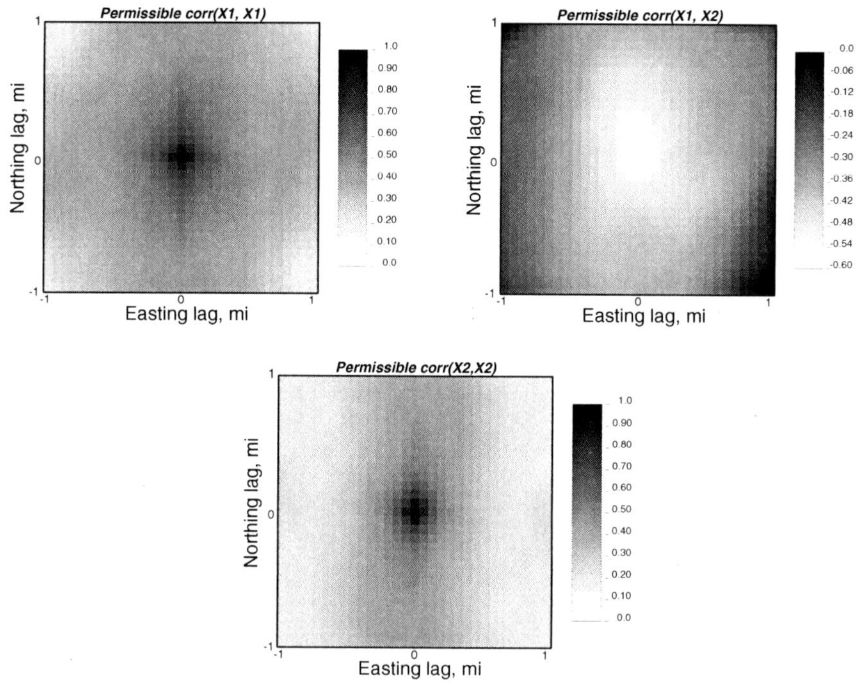

Figure 7.11: Permissible correlogram maps for alr transformation ($h\text{w} = 0$, $ns = 0$).

The following step is to adjust the correlogram maps to permissible correlograms, which was done using program `multsmth`. The program requires the setting of two parameters that have a significant impact on the results. In the absence of any criteria for determining parameters, the user is forced to proceed by trial and error relative to a reference; we selected the conventional, experimental points as a reference. In this sense the process is similar to the fitting of parameters for analytical models. The advantages in this case are that both parameters are integer numbers and that the program produces permissible correlograms for all maps in one run. Considering that the best solutions result when the values of such parameters are small, the combinations of possibilities are fairly limited. We started with

- Maximum half smoothing window size (hw): 0
- Minimum number of data for smoothing (ns): 0

the results of which are given in Figures 7.11 and 7.12 (A, B, C). Whereas the fitting is reasonable for the correlograms, the fitting of the cross-correlogram is poor.

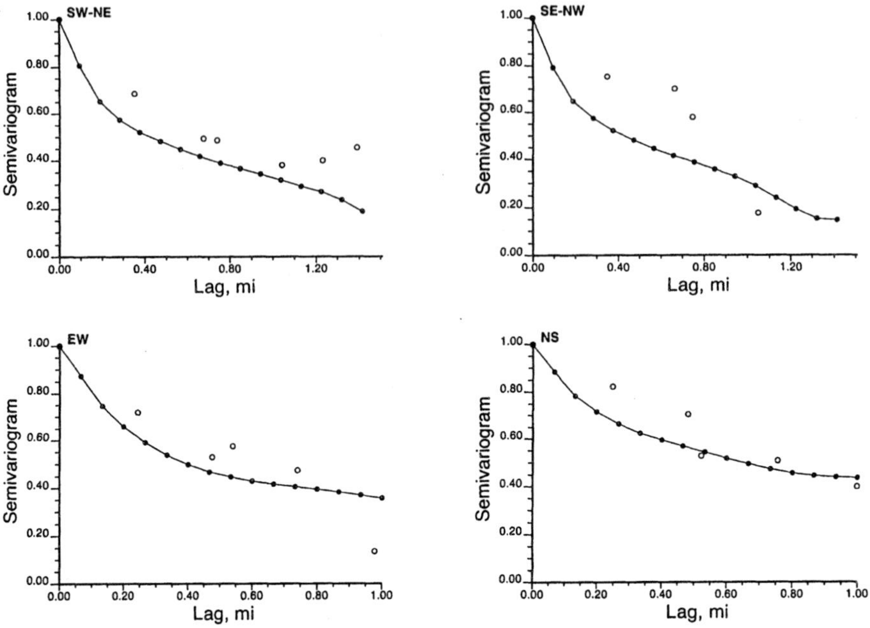

Figure 7.12: (A) Directional correlograms for X_1 for ($hw = 0$, $ns = 0$). Dots = values from maps in Fig. 7.11; open circles = conventional correlograms in Fig. 7.6 (A).

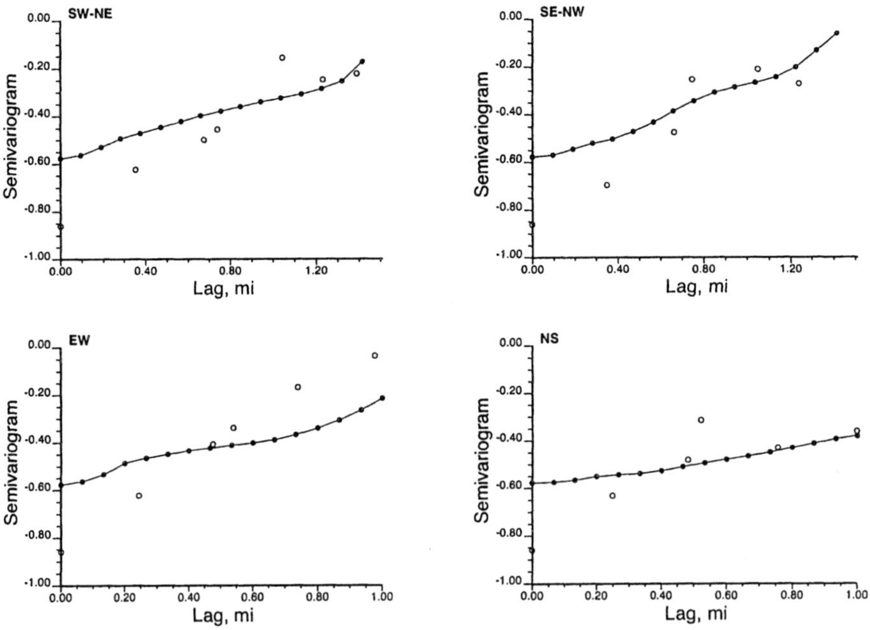

Figure 7.12: (B) Cross-correlograms between $X_1(\mathbf{u})$ and $X_2(\mathbf{u})$.

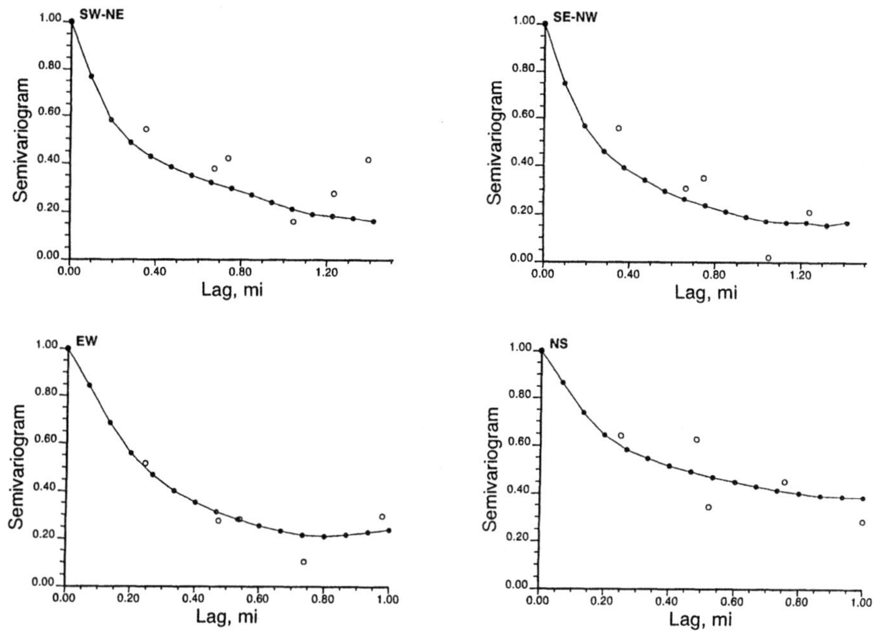

Figure 7.12: (C) Correlograms for $X_2(\mathbf{u})$.

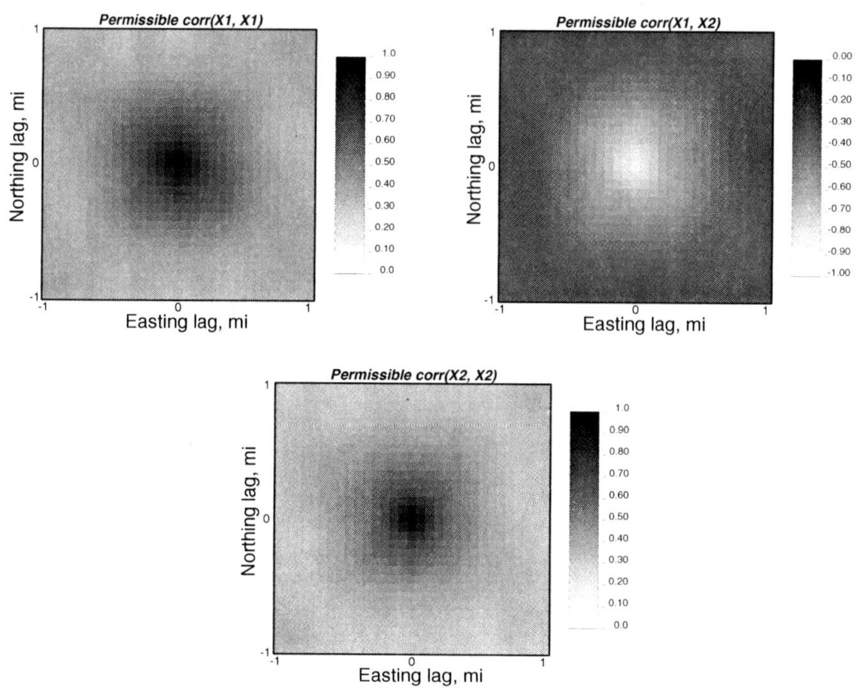

Figure 7.13: Permissible correlogram maps for alr transformation ($hw = 1, ns = 0$).

The preceding Figure 7.13 and Figure 7.14 (A, B, C) show the results for ($hw = 1$, $ns = 0$). Maps in Figures 7.11 and 7.13 are slightly smaller than the previous maps because of loss of one peripheral column and row in the processing by `multsmth`. Further increase in the parameters does not produce significantly better results. Hence the maps in Figure 7.13 are the best. For $hw < ns$, the cross-correlogram is even flatter than in Figure 7.12 (B). As the value of both parameters increases while keeping $hw \geq ns$, results are either extremely close to those for $(1, 0)$ when ns is much larger than hw or otherwise the maps are too peaked near the origin, even for the cross-correlogram. For example, maps for $(1, 0)$, $(5, 0)$, and $(5, 1)$ are essentially the same and $(5, 2)$, $(5, 3)$, $(5, 4)$, and $(5, 5)$ are too peaked.

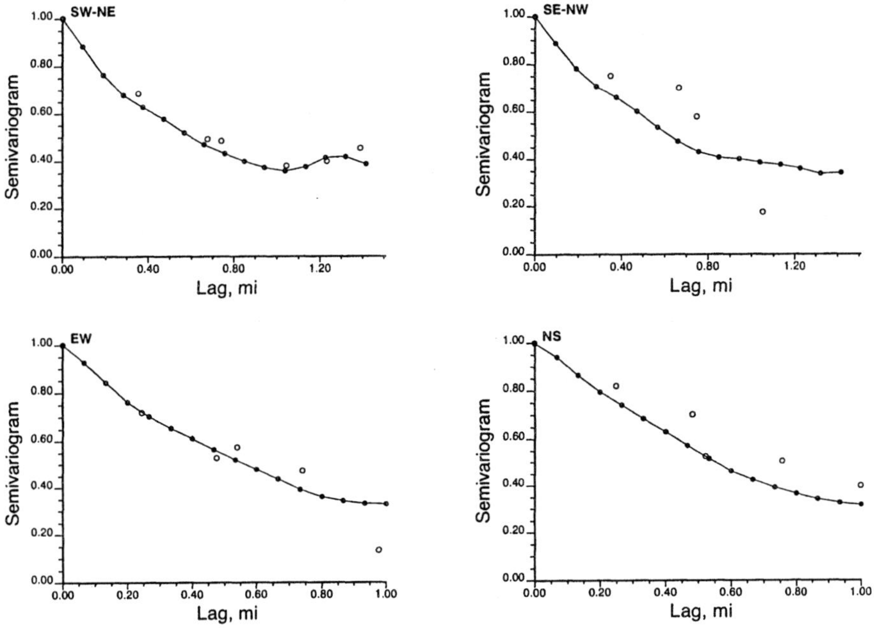

Figure 7.14: (A) Directional correlograms of X_1 for ($hw = 1$, $ns = 0$). Dots are values from maps in Fig. 7.13; open circles are conventional correlograms from Fig. 7.6 (A).

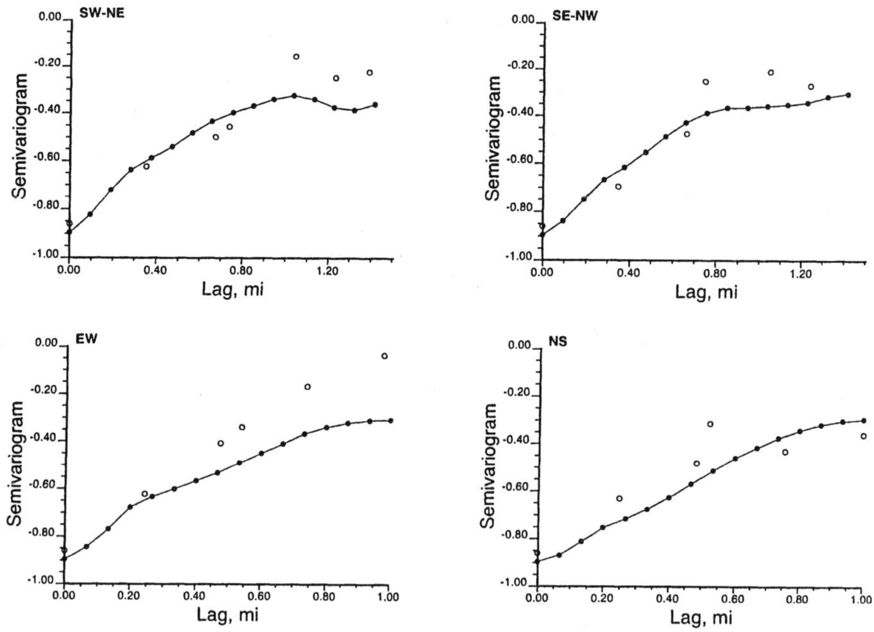

Figure 7.14: (B) Cross-correlograms between $X_1(\mathbf{u})$ and $X_2(\mathbf{u})$.

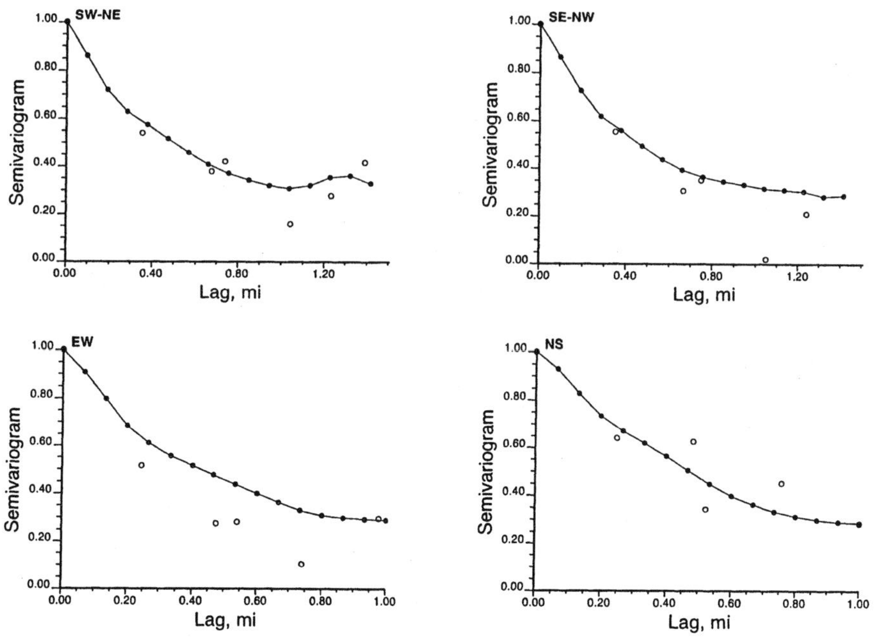

Figure 7.14: (C) Correlograms for $X_2(\mathbf{u})$.

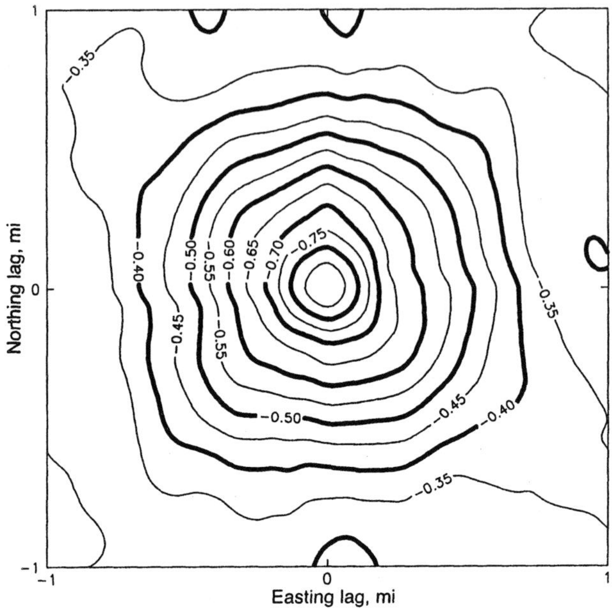

Figure 7.15: Contoured version of cross-correlation map in Fig. 7.13.

Figure 7.15 is an enlarged contour-map version of the estimated cross-correlogram in Figure 7.13, displayed again to show in more detail that

$$\hat{\rho}(X_1(\mathbf{u}), X_2(\mathbf{u})) \neq \hat{\rho}(X_1(-\mathbf{u}), X_2(-\mathbf{u})).$$

Finally, one needs to convert the correlograms to covariances, which under conditions of stationarity requires a simple scaling by the estimated standard deviations $\hat{\sigma}_k$

$$\hat{\text{Cov}}(X_1(\mathbf{u}), X_2(\mathbf{u})) = \hat{\sigma}_1 \hat{\sigma}_2 \hat{\rho}(X_1(\mathbf{u}), X_2(\mathbf{u})). \tag{7.5}$$

In the case of Lyons West, the conversion is more complex because of the presence of drift. This drift is revealed in semivariograms exponentially increasing above the value of a constant, minimum-value sill that can be observed along the trend-free direction (Olea 1999, pp. 75–77). Besides, in the multivariate case a drift exaggerates the correlation coefficients relative to those that would be obtained for the residuals after removal of the trend. As shown by Figure 7.16, the transformations

$$\begin{aligned} \hat{\text{Cov}}(X_1(\mathbf{u})) &= \hat{\sigma}_1^* \hat{\rho}(X_1(\mathbf{u})) - 0.039 \\ \hat{\text{Cov}}(X_1(\mathbf{u}), X_2(\mathbf{u})) &= 0.8\,(\hat{\sigma}_1 \hat{\sigma}_2 \hat{\rho}(X_1(\mathbf{u}), X_2(\mathbf{u})) - 0.016) \\ \hat{\text{Cov}}(X_2(\mathbf{u})) &= \hat{\sigma}_2^* \hat{\rho}(X_2(\mathbf{u})) - 0.009 \end{aligned} \tag{7.6}$$

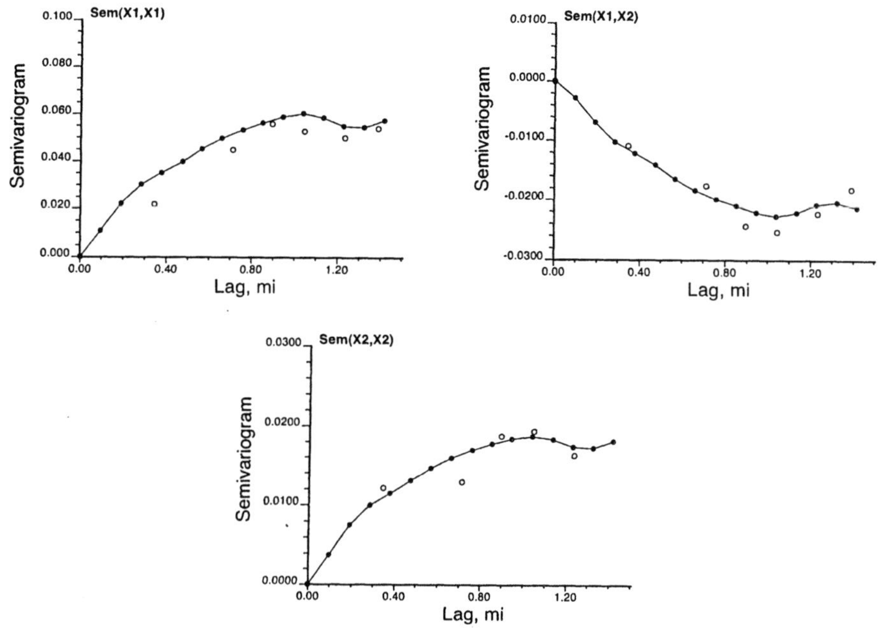

Figure 7.16: Semivariograms along the trend-free direction N 45 E. Dots denote values derived from correlograms in Fig. 7.13; open circles denote conventional experimental semivariogram estimates.

cause the semivariograms derived from the model to follow closely the experimental semivariogram along the trend-free direction N 45 E. The values of the standard deviation were taken from the sample and are $\hat{\sigma}_1 = 0.307$ and $\hat{\sigma}_2 = 0.165$. This transformation produced better cross-validation results than simply employing the transformation in equation (7.5) and replacing the variances by the sills along the trend-free directions.

7.3.3 Comparison of estimation methods

We employed the leaving-one-out technique to generate estimates of the alr transformed vector $\mathbf{X}(\mathbf{u}) = (X_1(\mathbf{u}), X_2(\mathbf{u}))'$ introduced in equation (7.3) at each sampling location $\mathbf{u}$ employing both ordinary kriging and ordinary cokriging. As shown in Figure 7.1, the almost regular pattern of the sampling is favorable for obtaining meaningful cross-validation results.

Estimates $\mathbf{X}^*(\mathbf{u}) = (X_1^*(\mathbf{u}), X_2^*(\mathbf{u}))'$ were agl back transformed [see equation (2.3)] to obtain estimates $\mathbf{Z}^*(\mathbf{u}) = (Z_o^*(\mathbf{u}), Z_w^*(\mathbf{u}), Z_r^*(\mathbf{u}))'$ in the simplex. Note that the estimates, displayed in Figure 7.17 by a plus sign (+), show less dispersion than the original data, represented by open circles (o). Note also the slight differences between estimates obtained using either ordinary kriging or ordinary cokriging.

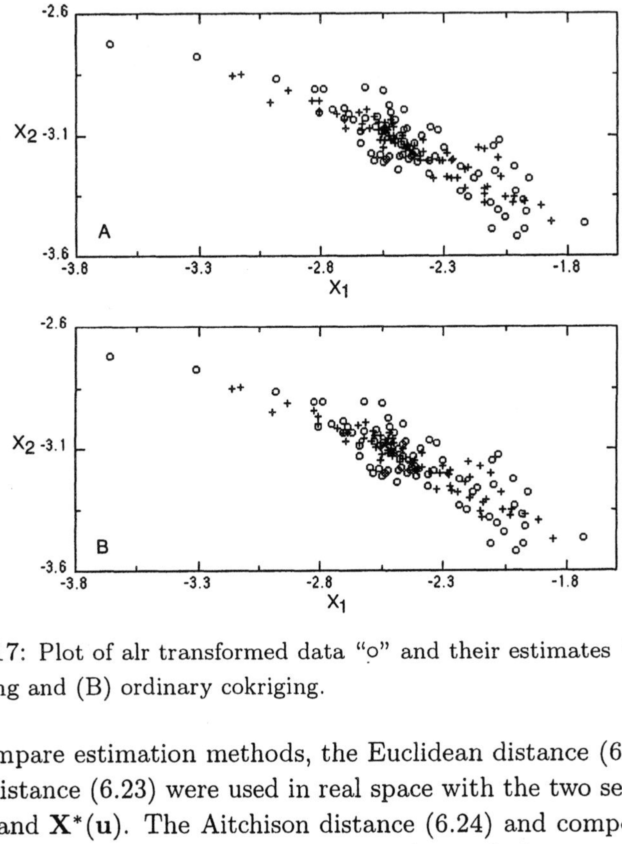

Figure 7.17: Plot of alr transformed data "o" and their estimates "+" by (A) ordinary kriging and (B) ordinary cokriging.

To compare estimation methods, the Euclidean distance (6.22) and Mahalanobis' distance (6.23) were used in real space with the two sets of data given by $\mathbf{X}(\mathbf{u})$ and $\mathbf{X}^*(\mathbf{u})$. The Aitchison distance (6.24) and compositional Mahalanobis' distance (6.25) were used on the simplex with the two sets of data given by $\mathbf{Z}(\mathbf{u})$ and $\mathbf{Z}^*(\mathbf{u})$. This was done both for cross-validation and for computing the STRESS. Univariate descriptive statistics of the vectors of distances and of the vectors of differences as defined in (6.21) are given in Tables 7.3, 7.4, and 7.5. A table corresponding to the cross-validation errors as measured by the compositional Mahalanobis' distance in the simplex is omitted, as it is identical to Table 7.4 [see equations (7.3) and (6.25)].

In Tables 7.3 and 7.5 it can be seen that both ordinary kriging and ordinary cokriging are for all practical purposes equivalent in the case of the Lyons West data set. When results are evaluated in real space using the Euclidean distance or in the simplex using Aitchison's distance, only minimal differences appear in the values of the maxima and of the parameters of skewness. This fact is

Table 7.3: Descriptive statistics for ordinary kriging and ordinary cokriging for the alr cross-validation errors as measured by Euclidean distance in the real space of the transforms and by the differential alr cross-validation errors.

Technique	Ordinary kriging $\delta_e(\mathbf{X}(\mathbf{u}), \mathbf{X}^*_{\text{ok}}(\mathbf{u}))$	Ordinary cokriging $\delta_e(\mathbf{X}(\mathbf{u}), \mathbf{X}^*_{\text{cok}}(\mathbf{u}))$	$\Delta_1(\mathbf{X}^*_{\text{ok}}, \mathbf{X}^*_{\text{cok}})$
Minimum	0.0400	0.0200	-0.0200
1st quartile (Q_1)	0.0700	0.0750	-0.0100
Median (Q_2)	0.1500	0.1400	0.0000
Mean	0.1571	0.1580	-0.0009
3rd quartile (Q_3)	0.2100	0.2200	0.0000
Maximum	0.5500	0.5500	0.0400
Std. deviation	0.0972	0.0985	0.0099
$(Q_3 - Q_1)/2$	0.0700	0.1000	0.0050
Skewness	1.2007	1.1604	1.0008

Table 7.4: Descriptive statistics for ordinary kriging and ordinary cokriging for the alr cross-validation errors as measured by Mahalanobis' distance in the real space of the transforms and by the differential cross-validation errors. The table is the same for the compositional Mahalanobis' distance in the simplex.

Technique	Ordinary kriging $\delta_m(\mathbf{X}(\mathbf{u}),\mathbf{X}^*_{\text{ok}}(\mathbf{u}))$	Ordinary cokriging $\delta_m(\mathbf{X}(\mathbf{u}),\mathbf{X}^*_{\text{cok}}(\mathbf{u}))$	$\Delta_2(\mathbf{X}^*_{\text{ok}},\mathbf{X}^*_{\text{cok}})$
Minimum	0.2240	0.1550	-0.2780
1st quartile (Q_1)	0.4925	0.4920	-0.0405
Median (Q_2)	0.8955	0.8605	-0.0060
Mean	0.9913	0.9912	0.0000
3rd quartile (Q_3)	1.3770	1.3685	0.0435
Maximum	2.9170	2.7500	0.2310
Std. deviation	0.5935	0.6064	0.0872
$(Q_3 - Q_1)/2$	0.4423	0.4383	0.0420
Skewness	0.9616	0.9018	-0.1077

Table 7.5: Descriptive statistics for compositional cross-validation errors derived from back transformation of alr transforms estimated by ordinary kriging and ordinary cokriging, as measured by Aitchison's distance in the simplex and by the differential cross-validation errors.

Technique	Ordinary kriging $\delta_a(\mathbf{Z}(\mathbf{u}), \mathbf{Z}^*_{\text{ok}}(\mathbf{u}))$	Ordinary cokriging $\delta_a(\mathbf{Z}(\mathbf{u}), \mathbf{Z}^*_{\text{cok}}(\mathbf{u}))$	$\Delta_3(\mathbf{Z}^*_{\text{ok}}, \mathbf{Z}^*_{\text{cok}})$
Minimum	0.0210	0.0150	-0.0150
1st quartile (Q_1)	0.0655	0.0615	-0.0045
Median (Q_2)	0.1320	0.1290	-0.0010
Mean	0.1433	0.1433	0.0000
3rd quartile (Q_3)	0.1960	0.1970	0.0040
Maximum	0.4980	0.4960	0.0390
Std. deviation	0.0949	0.0955	0.0076
$(Q_3 - Q_1)/2$	0.0653	0.0678	0.0043
Skewness	1.1717	1.1500	0.9635

corroborated by the values of the differential cross-validation errors,

$$\begin{aligned}
\Delta_1(\mathbf{X}^*_{\text{ok}}, \mathbf{X}^*_{\text{cok}}) &= \delta_e(\mathbf{X}(\mathbf{u}), \mathbf{X}^*_{\text{ok}}(\mathbf{u})) - \delta_e(\mathbf{X}(\mathbf{u}), \mathbf{X}^*_{\text{cok}}(\mathbf{u})) \\
\Delta_3(\mathbf{Z}^*_{\text{ok}}, \mathbf{Z}^*_{\text{cok}}) &= \delta_a(\mathbf{Z}(\mathbf{u}), \mathbf{Z}^*_{\text{ok}}(\mathbf{u})) - \delta_a(\mathbf{Z}(\mathbf{u}), \mathbf{Z}^*_{\text{cok}}(\mathbf{u})),
\end{aligned}$$

which show a zero median, a zero mean and nearly no dispersion. Larger differences appear for the Mahalanobis' distance in real space or, equivalently, for the compositional Mahalanobis' distance in the simplex. We see in Table 7.4 that ordinary kriging performs better if the (arithmetic) mean of the distances is considered as an appropriate representative measure of central tendency, while ordinary cokriging performs better if the median is considered to be more representative. Recall that distances are nonnegative by definition, and that the smaller the value, the better the estimate. Distances for ordinary cokriging show more or less the same dispersion as distances for ordinary kriging, as both the interquartile range and the standard deviation differ only in the second decimal. Neither method is favored above the other. The values of the differential cross-validation error,

$$\Delta_2(\mathbf{X}^*_{\text{ok}}, \mathbf{X}^*_{\text{cok}}) = \delta_m(\mathbf{X}(\mathbf{u}), \mathbf{X}^*_{\text{ok}}(\mathbf{u})) - \delta_m(\mathbf{X}(\mathbf{u}), \mathbf{X}^*_{\text{cok}}(\mathbf{u})),$$

show a large negative minimum, a negative mean and negative skewness, facts that could be interpreted as unsatisfactory cokriging results. But the median is very close to zero, indicating that in 50% of the cases, ordinary cokriging performs better or equally as well as ordinary kriging.

We plotted the three pairs of distances; graphical representations of the results using Q-Q plots are shown in Figures 7.18, 7.19, and 7.20. Figures 7.18 and 7.20 show that the segmented line is very close to the diagonal, thus confirming that, with this data set, both methods are identical for practical purposes. Figure 7.19, on the contrary, shows a clear tendency of ordinary kriging to perform better, although this tendency is largely influenced by some badly estimated cases, while the smaller distances play in favor of ordinary cokriging.

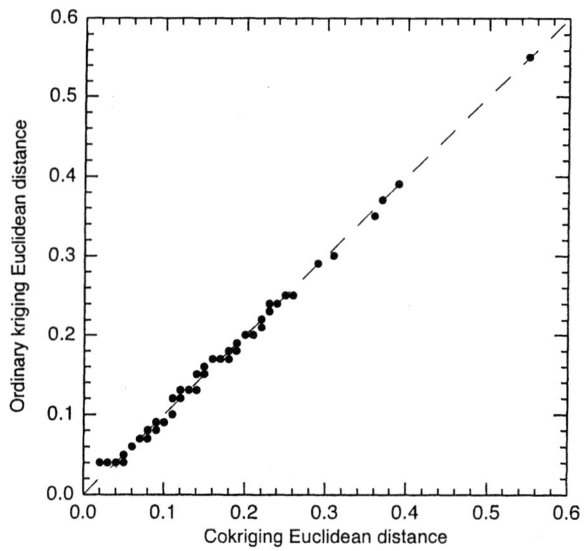

Figure 7.18: Q-Q plot for ordinary kriging and ordinary cokriging cross-validation errors for alr transformed values as measured by Euclidean distance in real space.

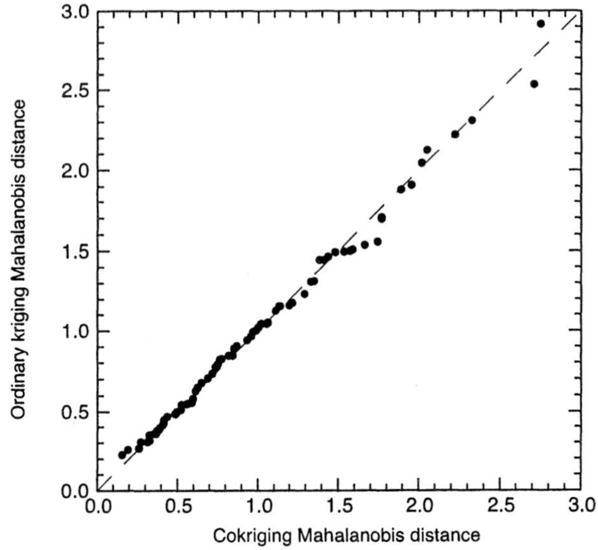

Figure 7.19: Q-Q plot for ordinary kriging and ordinary cokriging cross-validation errors for alr transformed values as measured by Mahalanobis' distance in real space. The plot is the same for the compositional Mahalanobis' distance in the simplex.

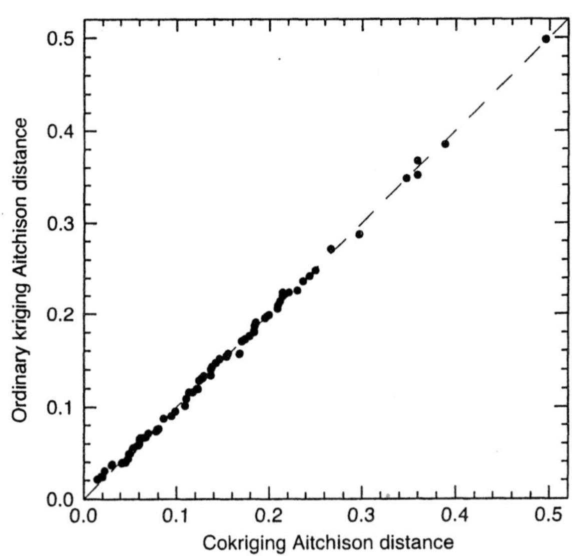

Figure 7.20: Q-Q plot for ordinary kriging and ordinary cokriging cross-validation errors for compositional variables as measured by Aitchison distance in the simplex.

It is difficult to come up with a conclusive result using distances and differential cross-validation errors following the previous reasoning, although the analysis is much easier compared to separately evaluating estimation errors for each component. Table 7.6 gives the STRESS using the different distances considered in each space. As expected from previous results, there is very small STRESS between estimates measured using either Euclidean distance in real space or Aitchison's distance in the simplex. We may consider these methods to be indistinguishable in this case.

The STRESS when comparing estimates with observed data for these two distances is less positive. Performance in general is not very good. The situation is similar if Mahalanobis' distance is used; the original dispersion of the samples is poorly reproduced by the estimates. The difference in the STRESS value in favor of ordinary kriging in all cases cannot be considered as significant. The small differences could be induced by numerical rounding errors. Nevertheless, it is in agreement with the tendency shown in Figure 7.19, which can be interpreted as representing a better performance by ordinary cokriging when both methods give good estimations, and a worse performance by ordinary cokriging when both estimations are poor.

Table 7.6: Performance of estimates for transforms and composition in the alr method as measured by STRESS. δ_e denotes Euclidean distance, δ_m Mahalanobis' distance, δ_a Aitchison's distance, δ_c compositional Mahalanobis' distance; ok is ordinary kriging, and cok is ordinary cokriging.

	Real space transform		Simplex composition	
	δ_e	δ_m	δ_a	δ_c
Data versus ok estimates	0.4347	0.5119	0.4386	0.5119
Data versus cok estimates	0.4364	0.5138	0.4399	0.5138
ok estimates versus cok estimates	0.0324	0.0729	0.0293	0.0729

7.4 The basis method

The basis method is derived using definitions and properties introduced in Section 2.2. It eliminates some of the peculiarities of compositional data by multiplying every composition by a size attribute.

7.4.1 The basis for Lyons West

Applying the basis method to the Lyons West data shown in Figure 7.1, we obtained

$$W_i(\mathbf{u}) = t(\mathbf{u})Z_i(\mathbf{u}), \quad i = o, w, r$$

for the saturated thickness, $t(\mathbf{u})$. This corresponds to the accumulation concept of mining geostatistics (Journel and Huijbregts 1978, p. 199). While compositional data measure relative abundance of each attribute, a regionalized basis denotes the total existing amount. As discussed in Section 2.2, the basis transformation removes the constant-sum constraint. Size is assumed to be a nonconstant, positive regionalized variable in itself, thus eliminating the linear dependence among components, although the nonnegative nature of the observations still remains. The basis method opens up the possibility of using either kriging or cokriging as an approximate estimation of the basis. Remember that these estimation methods assume that the sample space is unconstrained from minus infinity to plus infinity and, therefore, we have at best a reasonable approximation, and never an optimal model. Back transformation to the simplex of the estimated basis data is fairly straightforward. All that we need to do is divide each component $W_i^*(\mathbf{u})$, $i = o, w, r$, by the estimate of its size, $t^*(\mathbf{u})$, and multiply the result by the constant-sum C in order to assure that the estimates add up to C, the constant, which in our case is 100:

$$\begin{aligned} t^*(\mathbf{u}) &= W_o^*(\mathbf{u}) + W_w^*(\mathbf{u}) + W_r^*(\mathbf{u}) \\ Z_i^*(\mathbf{u}) &= C\frac{W_i^*(\mathbf{u})}{t^*(\mathbf{u})}, \quad i = o, w, r\,. \end{aligned}$$

7.4.2 Modeling of covariances

Preparation of permissible covariance maps is completely analogous to the case of the alr transforms in Section 7.3.2, except that the graphical display is more voluminous because we now have three variables, hence three covariances and three cross-covariances.

Figure 7.21 displays the set of optimal correlogram maps which, in this instance, occur when ($hw = 2$, $ns = 0$). The vector of standard deviations necessary for scaling the covariance function is $\hat{\sigma} = (1.02,\ 0.26,\ 7.26)'$.

7.4.3 The kriging and cokriging of the basis

In the particular case of the Lyons West study, we did not obtain negative estimates. As explained in Section 7.2, this result is dependent on the smoothing effect that characterizes both estimation methods, confirmed by the smaller

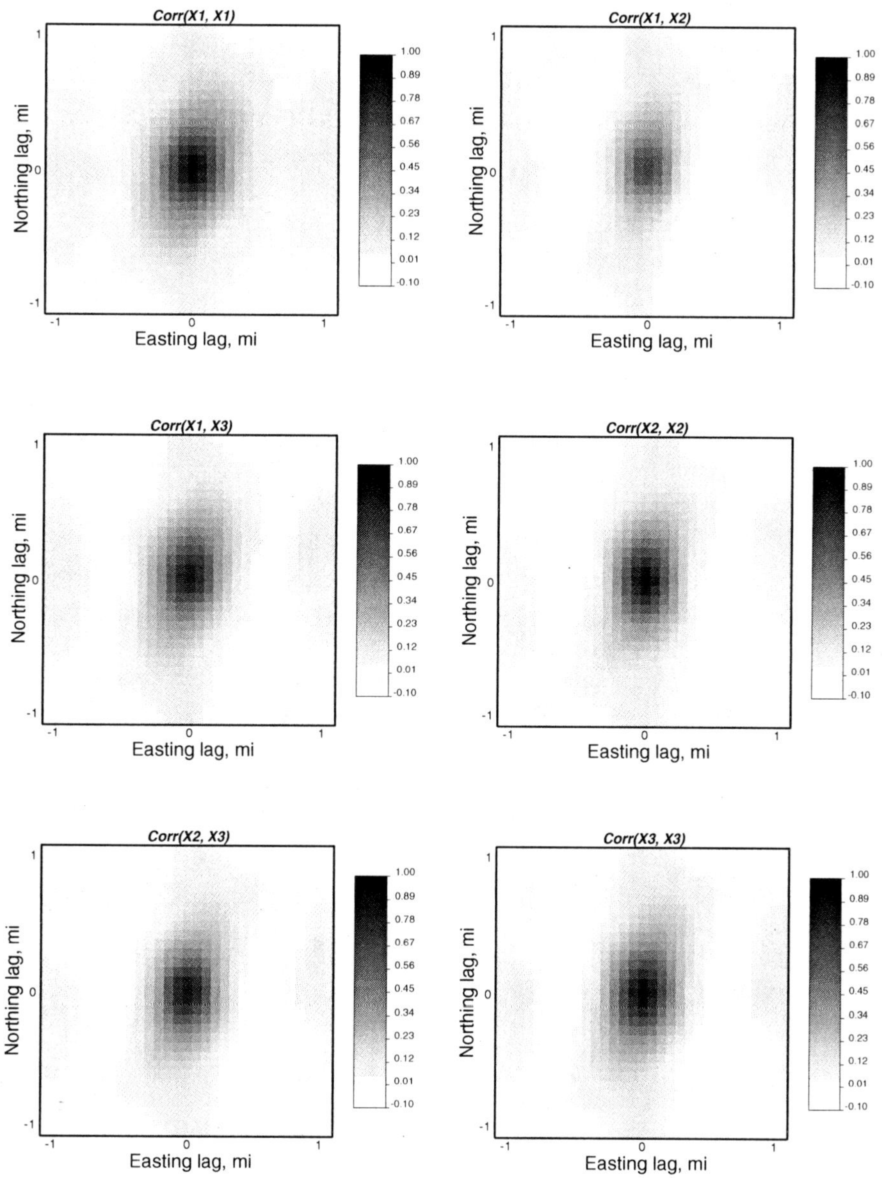

Figure 7.21: Best permissible correlogram maps for the basis transformation.

dispersion of estimated values as compared to observations shown in Figure 7.22. Note that the data set is located almost perfectly along a straight line; this observation will be important in the results we will discuss later.

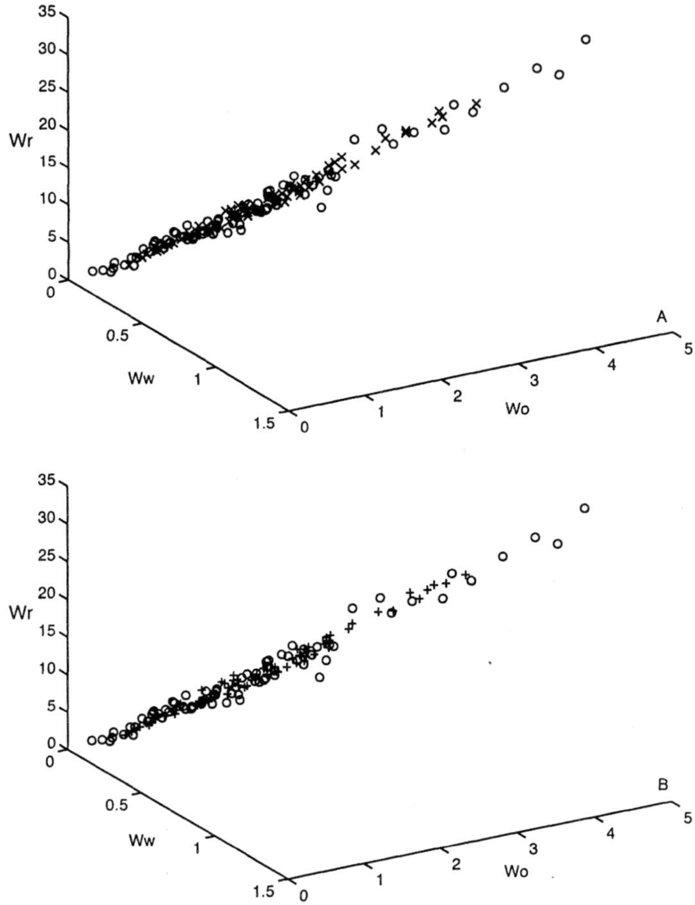

Figure 7.22: Dispersion of data "o" and smoothing of estimates "×': (A) Original observations $\mathbf{W}(\mathbf{u})$ and estimates by ordinary kriging, $\mathbf{W}^*_{\mathrm{ok}}(\mathbf{u})$; (B) Original observations and estimates by ordinary cokriging, $\mathbf{W}^*_{\mathrm{cok}}(\mathbf{u})$.

Let us now consider the differences using Euclidean distances between observations and estimated vectors for each of the two estimation methods in the basis space, which is the positive orthant of three-dimensional real space. The two vectors of the 76 estimation errors for ordinary kriging and ordinary cokriging, denoted by $\delta_e(\mathbf{W}(\mathbf{u}), \mathbf{W}^*_{\mathsf{ok}}(\mathbf{u}))$ and $\delta_e(\mathbf{W}(\mathbf{u}), \mathbf{W}^*_{\mathsf{cok}}(\mathbf{u}))$, allow us to define a third vector

$$\Delta_1(\mathbf{W}^*_{\mathsf{ok}}, \mathbf{W}^*_{\mathsf{cok}}) = \delta_e(\mathbf{W}(\mathbf{u}), \mathbf{W}^*_{\mathsf{ok}}(\mathbf{u})) - \delta_e(\mathbf{W}(\mathbf{u}), \mathbf{W}^*_{\mathsf{cok}}(\mathbf{u})).$$

Results are shown in Table 7.7. Note that distance $\delta_e(\mathbf{W}(\mathbf{u}), \mathbf{W}^*_{\mathsf{cok}}(\mathbf{u}))$ is smaller, on average, than $\delta_e(\mathbf{W}(\mathbf{u}), \mathbf{W}^*_{\mathsf{ok}}(\mathbf{u}))$, implying that in this case ordinary cokriging performs on average better than ordinary kriging.

Table 7.7: Descriptive statistics for ordinary kriging and ordinary cokriging for the basis cross-validation errors as measured by Euclidean distance in the real space of the transforms and by the differential basis cross-validation errors.

	Ordinary kriging $\delta_e(\mathbf{W}(\mathbf{u}),\mathbf{W}^*_{\mathsf{ok}}(\mathbf{u}))$	Ordinary cokriging $\delta_e(\mathbf{W}(\mathbf{u}),\mathbf{W}^*_{\mathsf{cok}}(\mathbf{u}))$	$\Delta_1(\mathbf{W}^*_{\mathsf{ok}},\mathbf{W}^*_{\mathsf{cok}})$
Minimum	0.0700	0.0500	-0.8300
1st quartile (Q_1)	0.6700	0.6100	-0.0550
Median (Q_2)	1.7050	1.5800	0.0050
Mean	2.8075	2.7779	0.0296
3rd quartile (Q_3)	4.7450	4.5600	0.1150
Maximum	15.6500	15.8400	0.8400
Std. deviation	2.9219	2.9286	0.2578
$(Q_3 - Q_1)/2$	2.0375	1.9750	0.0850
Skewness	1.7091	1.7751	0.1998

We can repeat this procedure for Mahalanobis' distances in the basis space; we then obtain the vectors $\delta_m(\mathbf{W}(\mathbf{u}), \mathbf{W}^*_{\mathsf{ok}}(\mathbf{u}))$, $\delta_m(\mathbf{W}(\mathbf{u}), \mathbf{W}^*_{\mathsf{cok}}(\mathbf{u}))$ and

$$\Delta_2(\mathbf{W}^*_{\mathsf{ok}}, \mathbf{W}^*_{\mathsf{cok}}) = \delta_m(\mathbf{W}(\mathbf{u}), \mathbf{W}^*_{\mathsf{ok}}(\mathbf{u})) - \delta_m(\mathbf{W}(\mathbf{u}), \mathbf{W}^*_{\mathsf{cok}}(\mathbf{u})).$$

In this instance, as shown in Table 7.8, the fact that the mean is negative indicates that, on average, Mahalanobis' distance $\delta_m(\mathbf{W}(\mathbf{u}), \mathbf{W}^*_{\mathsf{ok}}(\mathbf{u}))$ is smaller than $\delta_m(\mathbf{W}(\mathbf{u}), \mathbf{W}^*_{\mathsf{cok}}(\mathbf{u}))$, implying that using Mahalanobis' distance as a criterion, we would consider ordinary kriging to be better than ordinary cokriging. Differing results for different criteria, although not unusual in statistics, require an explanation, which in this case is probably related to the fact that both estimation methods—ordinary kriging and ordinary cokriging—are based on the minimization of Euclidean error. Nevertheless, further study is required to confirm this hypothesis.

Table 7.8: Descriptive statistics for ordinary kriging and ordinary cokriging for the basis cross-validation errors as measured by Mahalanobis' distance in the real space of the transforms and by the differential basis cross-validation errors.

	Ordinary kriging $\delta_m(\mathbf{W}(\mathbf{u}),\mathbf{W}^*_{\text{ok}}(\mathbf{u}))$	Ordinary cokriging $\delta_m(\mathbf{W}(\mathbf{u}),\mathbf{W}^*_{\text{cok}}(\mathbf{u}))$	$\Delta_2(\mathbf{W}^*_{\text{ok}},\mathbf{W}^*_{\text{cok}})$
Minimum	0.0450	0.0710	-2.7930
1st quartile (Q_1)	0.5155	0.6515	-0.2870
Median (Q_2)	0.9720	1.5555	-0.1035
Mean	1.1425	1.3387	-0.1962
3rd quartile (Q_3)	1.6565	1.7305	0.0200
Maximum	3.6510	5.2500	0.3770
Std. deviation	0.7601	0.9762	0.4856
$(Q3 - Q1)/2$	0.5705	0.5395	0.2435
Skewness	0.9206	1.5635	-3.5074

Following our scheme of comparing different approaches (Section 6.7), we computed the STRESS for the three data sets in the basis space: original observations $\mathbf{W}(\mathbf{u})$; estimates by ordinary kriging $\mathbf{W}^*_{\text{ok}}(\mathbf{u})$; and estimates by ordinary cokriging $\mathbf{W}^*_{\text{cok}}(\mathbf{u})$, using the Euclidean distance and Mahalanobis' distance separately.

For Euclidean distance, STRESS for the combination $\mathbf{W}(\mathbf{u})$ versus $\mathbf{W}^*_{\text{ok}}(\mathbf{u})$ is slightly larger than the STRESS for $\mathbf{W}(\mathbf{u})$ versus $\mathbf{W}^*_{\text{cok}}(\mathbf{u})$, although both are quite large, an indication that the estimates are not very close to the observed values. Nevertheless, the fact that the STRESS for ordinary cokriging is smaller and the STRESS $\mathbf{W}^*_{\text{ok}}(\mathbf{u})$ versus $\mathbf{W}^*_{\text{cok}}(\mathbf{u})$ is close to zero shows that ordinary cokriging is equivalent to ordinary kriging of the basis and that there is almost no difference between these estimation results when Euclidean distance is used to measure performance. The reverse is the case when Mahalanobis' distance is used, wherein ordinary kriging performs better and there is a clear difference between the two estimation methods. Table 7.9 summarizes results. A comparison of distances obtained from cross-validation both in the transformed basis space and in the simplex of the compositional data is displayed by Q-Q plots in Figure 7.23. The main diagonal is the line of indifference in the quality of the estimation: (A) has been obtained using Euclidean distance in the basis space for ordinary kriging and ordinary cokriging; (B) results from using the corresponding Mahalanobis' distance in the basis space, again for ordinary kriging and ordinary cokriging; and (C) and (D) correspond to the Aitchison distance and the compositional Mahalanobis' distance.

Table 7.9: Performance for cross-validation of the basis as measured by STRESS. Euclidean distance is denoted by δ_e, Mahalanobis' distance by δ_m, ordinary kriging by ok and ordinary cokriging by cok.

	Real space	Basis
	δ_e	δ_m
Data versus ok estimates	0.4642	0.4444
Data versus cok estimates	0.4619	0.4798
ok estimates versus cok estimates	0.0457	0.2829

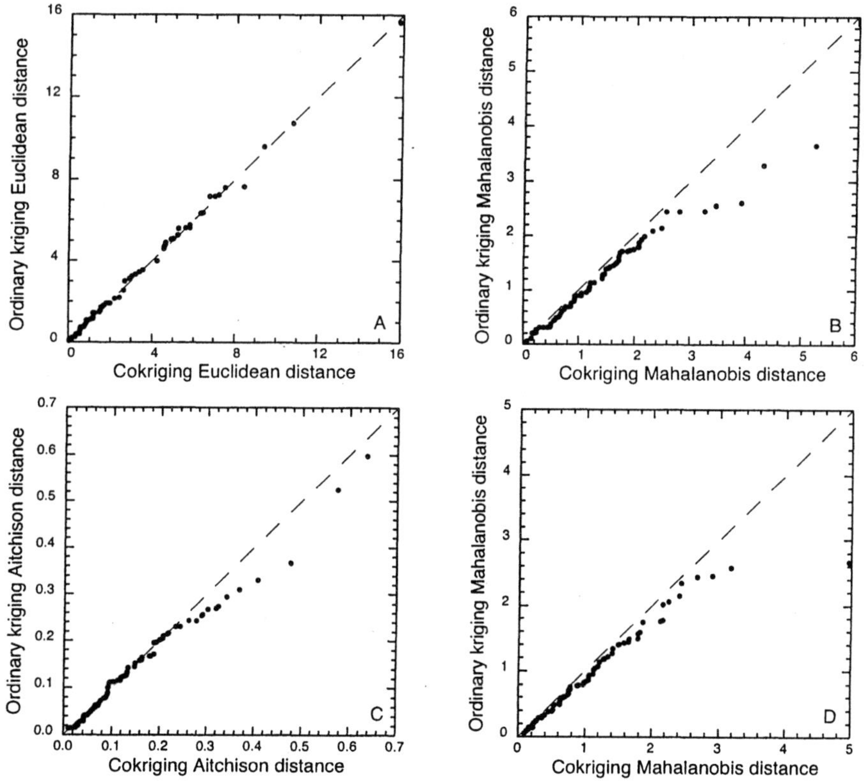

Figure 7.23: Evaluation of results for estimation by the basis method: Q-Q plots for (A) Euclidean distances between observations and estimates in the basis space; (B) Mahalanobis' distances between observations and estimates in the basis space; (C) Aitchison's distance between observations on the simplex and back transformed estimates; (D) the compositional Mahalanobis' distance between observations on the simplex and the back transformed estimates.

While ordinary cokriging appears to be a better estimator for the basis using Euclidean distance, it is interesting to note a reversal in the optimality produced by the back transformation which makes ordinary kriging the preferred estimation method for the composition in the simplex. In fact, let us turn our attention to the back transformed data set and compute the corresponding vectors of distances

$$\delta_a(\mathbf{Z}(\mathbf{u}), \mathbf{Z}^*_{\text{ok}}(\mathbf{u})) \quad \text{and} \quad \delta_a(\mathbf{Z}(\mathbf{u}), \mathbf{Z}^*_{\text{cok}}(\mathbf{u}))$$

and the vector of differences between both,

$$\Delta_3(\mathbf{Z}^*_{\text{ok}}, \mathbf{Z}^*_{\text{cok}}) = \delta_a(\mathbf{Z}(\mathbf{u}), \mathbf{Z}^*_{\text{ok}}(\mathbf{u})) - \delta_a(\mathbf{Z}(\mathbf{u}), \mathbf{Z}^*_{\text{cok}}(\mathbf{u})).$$

This procedure was repeated for the compositional Mahalanobis' distances

$$\delta_c(\mathbf{Z}(\mathbf{u}), \mathbf{Z}^*_{\text{ok}}(\mathbf{u})), \quad \delta_c(\mathbf{Z}(\mathbf{u}), \mathbf{Z}^*_{\text{cok}}(\mathbf{u}))$$

and

$$\Delta_4(\mathbf{Z}^*_{\text{ok}}, \mathbf{Z}^*_{\text{cok}}) = \delta_c(\mathbf{Z}(\mathbf{u}), \mathbf{Z}^*_{\text{ok}}(\mathbf{u})) - \delta_c(\mathbf{Z}(\mathbf{u}), \mathbf{Z}^*_{\text{cok}}(\mathbf{u})).$$

Results are shown in Tables 7.10 and 7.11. The mean is close to zero, but negative for Δ_3 and negative also—but larger—for Δ_4. The bulk of the distribution is negative and there is also negative skewness in both cases. Furthermore,

Table 7.10: Descriptive statistics for compositional cross-validation errors derived from back transformation of the basis estimated by ordinary kriging and ordinary cokriging, as measured by Aitchison's distance in the simplex and by the differential cross-validation errors.

	Ordinary kriging $\delta_a(\mathbf{Z}(\mathbf{u}), \mathbf{Z}^*_{\text{ok}}(\mathbf{u}))$	Ordinary cokriging $\delta_a(\mathbf{Z}(\mathbf{u}), \mathbf{Z}^*_{\text{cok}}(\mathbf{u}))$	$\Delta_3(\mathbf{Z}^*_{\text{ok}}, \mathbf{Z}^*_{\text{cok}})$
Minimum	0.0130	0.0100	-0.2470
1st quartile (Q_1)	0.0580	0.0650	-0.0350
Median (Q_2)	0.1150	0.1155	-0.0050
Mean	0.1404	0.1516	-0.0112
3rd quartile (Q_3)	0.2030	0.2030	0.0150
Maximum	0.5970	0.6350	0.1060
Std. deviation	0.1103	0.1246	0.0510
$(Q_3 - Q_1)/2$	0.0725	0.0690	0.0250
Skewness	1.6213	1.6865	-1.5336

Table 7.11: Descriptive statistics for compositional cross-validation errors derived from back transformation of the basis estimated by ordinary kriging and by ordinary cokriging, as measured by the compositional Mahalanobis' distance in the simplex and the differential cross-validation errors.

	Ordinary kriging $\delta_c(\mathbf{Z}(\mathbf{u}), \mathbf{Z}^*_{\text{ok}}(\mathbf{u}))$	Ordinary cokriging $\delta_c(\mathbf{Z}(\mathbf{u}), \mathbf{Z}^*_{\text{cok}}(\mathbf{u}))$	$\Delta_4(\mathbf{Z}^*_{\text{ok}}, \mathbf{Z}^*_{\text{cok}})$
Minimum	0.0550	0.0950	-2.5240
1st quartile (Q_1)	0.4780	0.5345	-0.2525
Median (Q_2)	0.7830	0.9400	-0.1030
Mean	0.9496	1.1058	-0.1563
3rd quartile (Q_3)	1.3630	1.4555	0.0240
Maximum	2.6590	4.9600	0.8700
Std. deviation	0.6500	0.8276	0.4467
$(Q_3 - Q_1)/2$	0.4425	0.4605	0.1383
Skewness	0.8604	1.7508	-1.9852

the STRESS of $\mathbf{Z}(\mathbf{u})$ versus $\mathbf{Z}^*_{\text{ok}}(\mathbf{u})$ is smaller than the STRESS for $\mathbf{Z}(\mathbf{u})$ versus $\mathbf{Z}^*_{\text{cok}}(\mathbf{u})$, while the STRESS for $\mathbf{Z}^*_{\text{ok}}(\mathbf{u})$ versus $\mathbf{Z}^*_{\text{cok}}(\mathbf{u})$ indicates a significant difference between both estimation methods. These results are summarized in Table 7.12. Although they may be surprising, there is an explanation that is associated to the strong linear dependency of the data set in the basis space which will be discussed in Section 7.4.4. Before we proceed, however, we would like to point out the fact that performance results obtained using Mahalanobis' distance in the basis space are consistent with results obtained for back transformed data. Again, ordinary kriging appears to be a better estimator, compared to ordinary cokriging.

Table 7.12: Performance for cross-validation of the basis, as measured by STRESS. Aitchison's distance is denoted by δ_a, δ_c denotes the compositional Mahalanobis' distance, ok is ordinary kriging and cok is ordinary cokriging.

	Simplex	Composition
	δ_a	δ_c
Data versus ok estimates	0.4519	0.5174
Data versus cok estimates	0.4805	0.5696
ok estimates versus cok estimates	0.1881	0.3385

7.4.4 Explanation of the reversal of the optimal method

Let us consider the 76 sampling sites provided by the Lyons West data and classify them into four groups: (1) BB, corresponding to sites where ordinary cokriging performed better in the basis space and in the simplex (in the sense that the error measured by the Euclidean distance in the basis space, and the error measured by the Aitchison distance in the simplex, is smaller in both cases for ordinary cokriging than for ordinary kriging; (2) BW, corresponding to those cases where ordinary cokriging performed better in the basis space and worse in the simplex; (3) WB, in the reverse case; and (4) WW, where ordinary cokriging performed worse in both spaces. To better describe the four groups from a graphical point of view, we will first illustrate the situation in 2-D (see Figure 7.24 A).

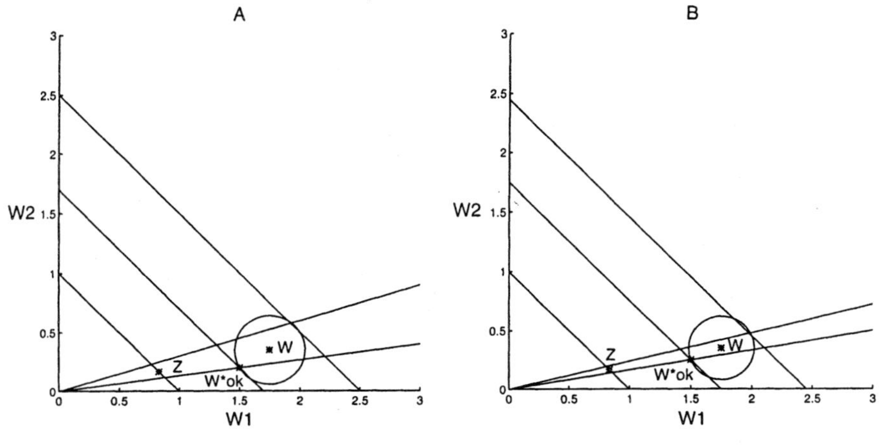

Figure 7.24: (A) Two-dimensional view of estimation space with observation $\mathbf{W}$, estimate using ordinary kriging $\mathbf{W}^*_{\text{ok}}$ and projection $\mathbf{Z}$. (B) Two-dimensional view of estimation space using ordinary kriging estimate when $\mathbf{W}^*_{\text{ok}}$ takes a position close to ray defined by sample $\mathbf{W}$.

Consider observation $\mathbf{W}$ and estimate $\mathbf{W}^*_{\text{ok}}$. Then, the situation *"the estimate using ordinary cokriging* $\mathbf{W}^*_{\text{cok}}$ *will be better than the estimate* $\mathbf{W}^*_{\text{ok}}$ *obtained by ordinary kriging in the basis space"* corresponds to those points lying inside the circumference because all points inside it will have a smaller Euclidean distance to $\mathbf{W}$ than $\mathbf{W}^*_{\text{ok}}$. The situation *"the back transformed ordinary cokriging estimate* $\mathbf{W}^*_{\text{cok}}$ *will be better than the back transformed ordinary kriging estimate* $\mathbf{W}^*_{\text{ok}}$ *in the simplex"* corresponds to all those points lying inside the section delimited by the radius through the origin and $\mathbf{W}^*_{\text{ok}}$, and the radius through the origin and $\mathbf{Z}'$, where $\mathbf{Z}'$ is obtained as follows: Consider the projection $\mathbf{Z}$ of $\mathbf{W}$, and $\mathbf{Z}^*_{\text{ok}}$ of $\mathbf{W}^*_{\text{ok}}$, on the simplex; let $\delta_a(\mathbf{Z}^*_{\text{ok}}, \mathbf{Z}^*_{\text{ok}})$ be the Aitchison distance between $\mathbf{Z}$ and $\mathbf{Z}^*_{\text{ok}}$. Then $\mathbf{Z}'$ is defined as that point on the simplex, different from $\mathbf{Z}^*_{\text{ok}}$, which has Aitchison distance $\delta_a(\mathbf{Z}^*_{\text{ok}}, \mathbf{Z}^*_{\text{ok}})$ from $\mathbf{Z}$. In fact, the projection on the simplex of any point lying inside this section will have a smaller Aitchison distance to $\mathbf{Z}$ than $\mathbf{Z}^*_{\text{ok}}$, as the Aitchison distance is monotonic. The intersection of the circumference and the section is BB, while that part of the circumference, which does not lie inside the section, is BW. WB corresponds to those points lying inside the section but outside the circumference, and the rest is WW. For each observation $\mathbf{W}$ of the original data set we will have an estimate $\mathbf{W}^*_{\text{ok}}$ obtained from ordinary kriging and we could build the corresponding graph. Then, the reversal phenomenon means that, on average, there are more cases where the ordinary cokriging estimate $\mathbf{W}^*_{\text{cok}}$ lies in the region BW, respectively WB, than in the regions BB or WW, as in the latter cases no reversal would appear.

In 3-D we have a similar situation, but with the circumference being a sphere, the section being a cone, the simplex being a pyramid, etc. Now, we can ask ourselves why it happens that on average more points go from better to worse than remain better or even go from worse to better. The answer lies in the previously mentioned fact that there is a quasi-linear behavior of our data in the basis space combined with the fact that we are minimizing the mean square error, which is actually the squared Euclidean distance. Our data have an extremely elongated position in the basis space (see Figure 7.22), as can be seen from a linear regression of the third component (total amount of rock) depending on the other two components through the origin, which accounts for about 98% of the original variability. The strong linear relationship, giving the strongest weight to the third component, implies that the original data lie on a nearly straight line on a ray through the origin, which is very close to the corresponding coordinate axis. Therefore, estimates using ordinary kriging will tend to take a position close to the ray defined by the sample and therefore defining a very small cone. See Figure 7.24 (B) for an illustration in 2-D.

On the other hand, the usage of ordinary cokriging will tend to position the estimate inside the sphere defined by the ordinary kriging estimate, but outside the cone, as a result of implicitly using the Euclidean distance in the estimation procedure. The result is the previously mentioned reversal. Using Mahalanobis' distance instead will lead to a very different picture: Equidistant points from the observation will lie on an ellipse in 2-D, and on a 3-D ellipsoid in 3-D. This ellipsoid having an elongated shape due to the mentioned characteristics of the sample, it will be similar to the section (cone in 3-D), and points will belong either to class BB or to class WW, as can be seen from descriptive statistics in Tables 7.7, 7.8, 7.10, and 7.11.

7.5 Last exercise

Figure 7.4 (A) presents optimistic results in terms of closure errors. It almost seems that in practice, despite all theoretical considerations, one can circumvent them. Here we will prove with a counterexample that such is not always the case.

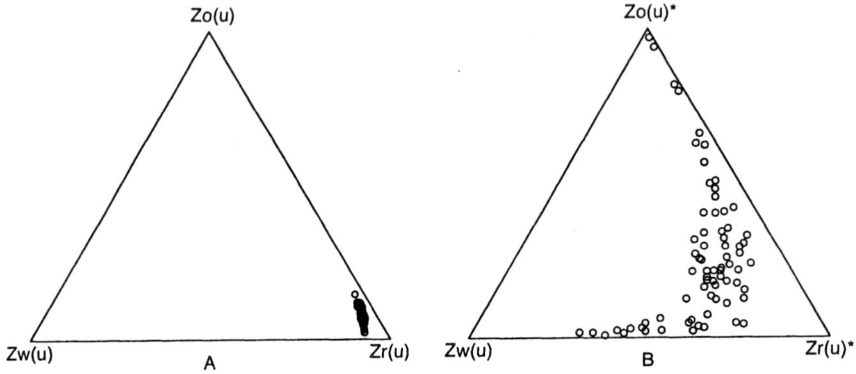

Figure 7.25: Ternary diagrams for Lyons West compositions. (A) Composition derived from the original data. (B) Composition modified to increase dispersion and curvature in the simplex.

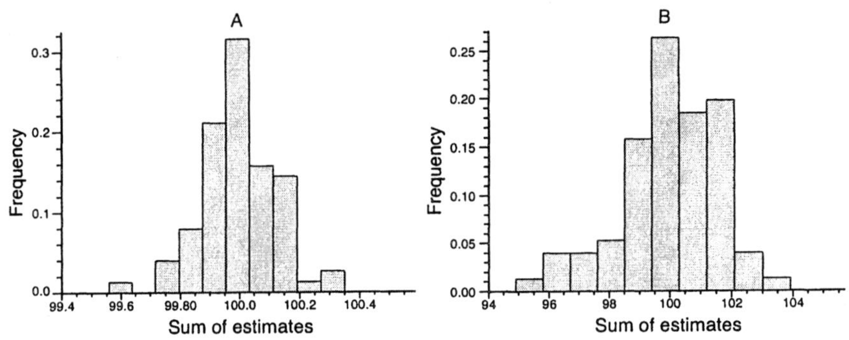

Figure 7.26: Histograms for cross-validation closure errors at the Lyons West field: (A) Original composition. (B) Modified composition.

Figure 7.25 (A) repeats the ternary diagram for the original data given in Figure 7.4 and displays (B) a new data set obtained after conveniently transforming the data to increase their dispersion and curvature in the simplex.

This transformation was achieved using perturbation and power transformation (Aitchison 1986).

By repeating the direct estimation that led to the closure errors in Figure 7.4 (A), we can use the transformed set to prepare another direct estimation. The old and new closure errors are illustrated in Figure 7.26. Errors in Figure 7.26 (B) are an order of magnitude larger than those in 7.26 (A), thus negating the proposition that compositional closure errors in direct estimation are always small—say, less than 1%.

7.6 Concluding comparisons

Because a completely general, theoretical approach is not possible, we have illustrated the advantages and disadvantages of the alr transformation and the basis method numerically by using a compositional data set of size 3 from the Lyons West oil field in Kansas, USA.

For the analysis we employed ordinary kriging and ordinary cokriging to generate cross-validation errors. To obtain diagnostics of performance, we calculated several statistics based on Euclidean, Mahalanobis' or Aitchison distances obtain from the multidimensional cross-validation errors.

There is no global method for estimating components of a regionalized composition simultaneously nor it is theoretically or practically sound to directly estimate each spatial component separately. The safest practice is to transform the data. Although the amount of improvement in results achieved by transformation methods compared to direct estimation varies, in the absence of criteria to decide when a transformation will produce significantly better results, the safest practice is always to use a transformation. Differences among transformation methods and estimation methods are of secondary importance relative to the improvements that can be achieved by using transformation methods instead of a direct approach.

In the special case of full sampling, the experience of others is confirmed by the Lyons West example. There are no significant improvements brought about by use of the more demanding modeling of ordinary cokriging rather than the simpler ordinary kriging. Under these circumstances, the parsimonious approach is to use ordinary kriging. In general, however, the safest practice is to employ cokriging, particularly for incomplete samplings.

As for the best transformation method, using STRESS as a measure of proximity of estimates to the observed vector observations, Tables 7.6 and 7.12 show that the alr method is slightly superior to the basis method regardless of the metric. Using Aitchison's distance, the alr method outperforms the basis method 0.4386 to 0.4519 if ordinary kriging is used and 0.4399 to 0.4805 when ordinary cokriging is used. The difference using Mahalanobis' distance is 0.5119 to 0.5171. Hence, in terms of transformation method, the slightly better performance of the alr method adds to its theoretical superiority in the sense that the domain of the alr transforms is real space while the domain of the basis method is restricted to the positive orthant.

Summary and prospects

If a multivariate regionalized data set is considered in which the components constitute an r-composition or are a subvector of an r-composition, then the approach presented here for spatial-structure analysis and estimation of r-compositions can be applied. Multivariate normality of the alr-transformed data is an additional assumption.

The first steps in the analysis of any r-composition include:

1. verification of stationarity assumptions,
2. investigation of nonsymmetry in the spatial covariance structure, and
3. determination of significant spatial dependence between the variables that are being analyzed.

Use of spectral methods is the most convenient way to analyze spatial continuity by the permissible modeling of covariances and cross-covariances taking into account their mutual relationships. Whether the approach is consistent with knowledge about the natural phenomenon under study should be a primary consideration at all steps.

Cokriging can produce estimates for the r-composition at unsampled sites as well as a vector of cokriging estimation variances, all in the real space of the transform. Back transformation of the estimates together with confidence regions based on the estimation variances is always possible, leading to confidence regions in the simplex.

Results from the Lyons West example lead us to conclude that the methodology described in this monograph opens up new possibilities for interpretation of compositional data for help in answering geologic questions. A limitation of the method is the assumption of the additive logistic normal distribution. Of course it cannot be expected that all r-compositions will follow such a distribution. Applicability of the procedures described here to other distributions must be examined. Extensions for alternative distributions may be found in

Aitchison (1986). Certainly the development of multivariate nonparametric procedures would also be possible.

A second problem worth further attention is the unbiasedness and minimum variance properties assigned to the back transformed results. Use of transformed data to analyze samplings using the large amount of robust methods available for the multivariate normally distributed phenomena is certainly convenient. However it remains to be proved whether the optimality properties of such estimators in a minimum mean-square-error sense in real space using Euclidean distances are preserved after back transformation to the simplex, where the Euclidean distances do not apply.

Finally there is by extension the challenge not only of working with Aitchison distances in the simplex instead of Euclidean distances, but also defining covariances and cross-covariances in the simplex using Aitchison distances. The next step would be to use these covariances and the metric to formulate estimation methods in the simplex; this would allow us to define and solve estimation problems in the simplex, eliminating the need to employ transformations.

References

Abbaspour, K. C., R. Schulin, M. T. van Genuchten, and E. Schläppi (1998). An alternative to cokriging for situations with small sample sizes. *Mathematical Geology 30*(3), 259–274.

Aboufirassi, M. and M. A. Mariño (1984). Cokriging of aquifer transmissivities from field measurements of transmissivity and specific capacity. *Mathematical Geology 16*(1), 19–35.

Ahmed, S. and G. de Marsily (1989). Co-kriged estimates of transmissivities using jointly water level data. See Armstrong (1989), pp. 615–628.

Aitchison, J. (1981a). Distributions on the simplex for the analysis of neutrality. In C. Taillie, G. P. Patil, and B. A. Baldessari (Eds.), *Statistical Distributions in Scientific Work—Models, Structures, and Characterizations*, pp. 147–156. Dordrecht: D. Reidel Publishing Co.

Aitchison, J. (1981b). A new approach to null correlations of proportions. *Mathematical Geology 13*(2), 175–189.

Aitchison, J. (1982). The statistical analysis of compositional data (with discussion). *Journal of the Royal Statistical Society, Series B (Statistical Methodology) 44*(2), 139–177.

Aitchison, J. (1983). Principal component analysis of compositional data. *Biometrika 70*(1), 57–65.

Aitchison, J. (1984a). The statistical analysis of geochemical compositions. *Mathematical Geology 16*(6), 531–564.

Aitchison, J. (1984b). Reducing the dimensionality of compositional data sets. *Mathematical Geology 16*(6), 617–636.

Aitchison, J. (1985). A general class of distributions on the simplex. *Journal of the Royal Statistical Society, Series B (Statistical Methodology) 47*(1), 136–146.

Aitchison, J. (1986). *The Statistical Analysis of Compositional Data*. Monographs on Statistics and Applied Probability. London: Chapman & Hall Ltd. 416 p.

Aitchison, J. (1989). Measures of location of compositional data sets. *Mathematical Geology 21*(7), 787–790.

Aitchison, J. (1990). Relative variation diagrams for describing patterns of compositional variability. *Mathematical Geology 22*(4), 487–511.

Aitchison, J. (1992). On criteria for measures of compositional difference. *Mathematical Geology 24*(4), 365–379.

Aitchison, J. (1997). The one-hour course in compositional data analysis or compositional data analysis is simple. See Pawlowsky-Glahn (1997), pp. 3–35.

Aitchison, J., C. Barceló-Vidal, A. J. Martín-Fernández, and V. Pawlowsky-Glahn (2000). Logratio analysis and compositional distance. *Mathematical Geology 32*(3), 271–275.

Aitchison, J. and J. A. C. Brown (1957). *The Lognormal Distribution.* Cambridge: Cambridge University Press. 176 p.

Aitchison, J. and S. M. Shen (1980). Logistic-normal distributions. Some properties and uses. *Biometrika 67*(2), 261–272.

Aitchison, J. and C. W. Thomas (1998). Differential perturbation processes: A tool for the study of compositional processes. See Buccianti, Nardi, and Potenza (1998), pp. 499–504.

Andrews, D. F., R. Gnanadesikan, and J. L. Warner (1971). Transformations of multivariate data. *Biometrics 27*, 825–840.

Armstrong, M. (Ed.) (1989). *Geostatistics*, Volume 1. Dordrecht: Kluwer Academic Publishers. 2 Vols., 1038 p.

Asli, M. and D. Marcotte (1995). Comparison of approaches to spatial estimation in a bivariate context. *Mathematical Geology 27*(5), 641–658.

Atkinson, P. M., R. Webster, and P. J. Curran (1992). Cokriging with ground-base radiometry. *Remote Sensing of Environment 41*(1), 45–60.

Azzalini, A. and A. Dalla Valle (1996). The multivariate skew-normal distribution. *Biometrika 83*(4), 715–726.

Bacon-Shone, J. (1992). Ranking methods for compositional data. *Applied Statistics 41*(3), 533–537.

Barceló-Vidal, C. (1996). *Mixturas de Datos Composicionales.* Ph. D. thesis, Universitat Politècnica de Catalunya, Barcelona. 261 p.

Barnes, R. J. (1991). The variogram sill and the sample variance. *Mathematical Geology 23*(4), 673–678.

Bonano, E. J., P. A. Davis, R. L. Bras, and P. K. Kitanidis (1989). Methodology for estimating groundwater travel times at a nuclear waste repository using a physically based geostatistical approach. See Armstrong (1989), pp. 673–684.

Bordessoule, J. L., C. Demange, and J. Rivoirard (1989). Using an orthogonal residual between ore and metal to estimate *in-situ* uranium resources. See Armstrong (1989), pp. 923–934.

Borgman, L. E. and R. B. Frahme (1976). A case study: Multivariate properties of bentonite in northern Wyoming. See Guarascio, David, and Huijbregts (1976), pp. 381–390.

Bourgault, G. and D. Marcotte (1991). Multivariable variogram and its application to the linear model of coregionalization. *Mathematical Geology 23*(7), 899–928.

Brunsdon, T. M. and T. M. F. Smith (1998). The time series analysis of compositional data. *Journal of Official Statistics 14*(3), 237–253.

Bryan, R. C. and F. Roghani (1982). Application of conventional and advanced methods to uranium reserve estimation and the development of a method to adjust for disequilibrium problems. In T. B. Johnson and R. J. Barnes (Eds.), *Application of Computers and Operations Research in the Mineral Industry,* Volume 17, pp. 109–120. New York: Society of Mining Engineers of the American Institute of Mining, Metallurgical and Petroleum Engineers (AIME).

Buccianti, A., G. Nardi, and R. Potenza (Eds.) (1998). *Proceedings of IAMG'98—The Fourth Annual Conference of the International Association for Mathematical Geology.* Naples: De Frede Editore. 2 Vols., 969 p.

Buccianti, A., V. Pawlowsky-Glahn, C. Barceló-Vidal, and E. Jarauta-Bragulat (1999). Visualization and modeling of natural trends in ternary diagrams: A geochemical case study. See Lippard, Næss, and Sinding-Larsen (1999), pp. 139–144.

Buccianti, A. and F. Rosso (1999). A new approach to the statistical analysis of compositional (closed) data with observations below the "detection limit". *Geoinformatica 3*, 17–31.

Buckles, R. S. (1965). Correlating and averaging connate water saturation data. *The Journal of Canadian Petroleum Technology 4*(1), 42–52.

Carr, J. R. and P. G. McCallister (1985). An application of cokriging for estimation of tripartite earthquake response spectra. *Mathematical Geology 17*(5), 527–545.

Carr, J. R. and D. E. Myers (1984). Application of the theory of regionalized variables to the spatial analysis of Landsat data. In *Proceedings of the Pecora 9 Spatial Information Technologies for Remote Sensing Today and Tomorrow*, pp. 55–61. Silver Spring, Md.: IEEE Computer Society Press.

Carr, J. R., D. E. Myers, and C. E. Glass (1985). Cokriging—A computer program. *Computers & Geosciences 11*(2), 111–127.

Chayes, F. (1962). Numerical correlation and petrographic variation. *The Journal of Geology 70*(4), 440–452.

Chayes, F. (1971). *Ratio Correlation.* Chicago: University of Chicago Press. 99 p.

Chilès, J.-P. and P. Delfiner (1999). *Geostatistics—Modeling Spatial Uncertainty* (2 ed.). Wiley Series in Probability and Statistics. New York: John Wiley & Sons. 695 p.

Christensen, R. (1991). *Linear models for Multivariate, Time Series and Spatial Data.* New York: Springer-Verlag. 317 p.

Clark, I. (1979). *Practical Geostatistics.* London: Applied Science Publishers. 129 p.

Clark, I., K. L. Basinger, and W. V. Harper (1989). MUCK—A novel approach to cokriging. In B. E. Buxton (Ed.), *Proceeding of the Conference on Geostatistical, Sensitivity, and Uncertainty Methods for Ground-Water Flow and Radionuclide Transport Modeling*, pp. 473–493. Columbus, Ohio: Battelle Press.

Cox, T. F. and M. A. A. Cox (1994). *Multidimensional Scaling,* Vol. 59 of *Monographs on Statistics and Applied Probability.* London: Chapman & Hall. xii+213 p. with 1 IBM-PC floppy disk (3.5 in. HD).

Cressie, N. (1991). *Statistics for Spatial Data.* New York: John Wiley & Sons. 900 p.

Creutin, J. D., H. Andrieu, and G. Delrieu (1989). Une simplification du cokrigeage appliquée á l'étalonnage d'images de télédétection en hydrométéorologie. See Armstrong (1989), pp. 761–772.

Daly, C., D. Jeulin, and C. Lajaunie (1989). Application of multivariate kriging to the processing of noisy images. See Armstrong (1989), pp. 749–760.

David, M. (1977). *Geostatistical Ore Reserve Estimation.* Series on Developments in Geomathematics, Volume 2. New York: Elsevier. 364 p.

Davis, B. M. (1987). Uses and abuses of cross-validation in geostatistics. *Mathematical Geology 19*(1), 241–248.

Davis, B. M. and K. A. Greenes (1983). Estimation using spatially distributed multivariate data: An example with coal quality. *Mathematical Geology 15*(2), 287–300.

Davis, P. J. and P. Rabinowitz (1975). *Methods of Numerical Integration.* New York: Academic Press. 459 p.

De Gruijter, J. J., D. J. J. Walvoort, and P. F. M. van Gaans (1997). Continuous soil maps—A fuzzy set approach to bridge the gap between aggregation levels of process and distribution models. *Geoderma 77*, 169–195.

Deutsch, C. and A. Journel (1998). *GSLIB—Geostatistical Software Library and User's Guide* (2 ed.). New York: Oxford University Press. 308 p. with 1 compact disk.

Diggle, P. J., J. A. Tawn, and R. A. Moyeed (1998). Model-based geostatistics (with discussion). *Journal of the Royal Statistical Society, Series C (Applied Statistics) 47*(3), 299–350.

Dowd, P. A. (1982). Lognormal kriging—The general case. *Mathematical Geology 14*(5), 475–499.

Dowd, P. A. (1989). Some observations on confidence intervals and kriging errors. See Armstrong (1989), pp. 861–874.

Doyen, P. M. (1988). Porosity from seismic data: A geostatistical approach. *Geophysics 53*(10), 1263–1275.

Ehm, A. E. (1965). Lyons West field: Kansas oil and gas fields. Technical report, Kansas Geological Society. pp. 146–156.

Fahrmeir, L. and A. Hamerle (1984). *Multivariate Statistische Verfahren.* Berlin: Walter de Gruyter. 796 p.

Feller, W. (1968). *An Introduction to Probability Theory and Its Applications* (3 ed.). New York: John Wiley & Sons. Vol. I, 501 p.

François-Bongarçon, D. (1981). Les coregionalisations, le cokrigeage. Technical Report C–86, École Nationale Supérieure des Mines de Paris, Centre de Geostatistique et de Morphologie Mathematique, Fontainebleau. 25 p.

Glass, C. E., J. R. Carr, H. M. Yang, and D. E. Myers (1988). Application of spatial statistics to analyzing multiple remote sensing data sets (STP 967). In A. I. Johnson and C. B. Pettersson (Eds.), *Geotechnical Applications of Remote Sensing and Remote Data Transmission*, pp. 138–150. Philadelphia: American Society for Testing and Materials.

Golub, G. H. and C. F. Van Loan (1989). *Matrix Computations.* Baltimore, Md.: The Johns Hopkins University Press. 642 p.

Goovaerts, P. (1994a). On a controversial method for modeling a coregionalization. *Mathematical Geology 26*(2), 197–204.

Goovaerts, P. (1994b). Study of spatial relationships between two sets of variables using multivariate geostatistics. *Geoderma 62*(1–3), 93–107.

Goovaerts, P. (1997). *Geostatistics for Natural Resources Evaluation.* Applied Geostatistics Series. New York: Oxford University Press. 483 p.

Goovaerts, P. and R. Webster (1994). Scale-dependent correlation between topsoil copper and cobalt concentrations in Scotland. *European Journal of Soil Science 45*, 79–95.

Goulard, M. (1989). Inference in a coregionalization model. See Armstrong (1989), pp. 397–408.

Goulard, M. and M. Voltz (1992). Linear coregionalization model: Tools for estimation and choice of cross-variogram matrix. *Mathematical Geology 24*(3), 269–286.

Grunsky, E. C. (1989). Spatial factor analysis: A technique to assess the spatial relationships of multivariate data. In F. P. Agterberg and G. F. Bonham-Carter

(Eds.), *Statistical Applications in the Earth Sciences*, pp. 329–347. Ottawa: Geological Survey of Canada.

Grunsky, E. C. and F. P. Agterberg (1988). Spatial and multivariate analysis of geochemical data from metavolcanic rocks in the Ben Nevis area, Ontario. *Mathematical Geology 20*(7), 825–861.

Grunsky, E. C. and F. P. Agterberg (1992). Spatial relationships of multivariate data. *Mathematical Geology 24*(6), 731–758.

Grunwald, G. K., A. E. Raftery, and P. Guttorp (1993). Time series of continuous proportions. *Journal of the Royal Statistical Society, Series B (Statistical Methodology) 55*(1), 103–116.

Guarascio, M. (1976). Improving the uranium deposit estimations (The Novazza Case). See Guarascio, David, and Huijbregts (1976), pp. 351–367.

Guarascio, M., M. David, and C. Huijbregts (Eds.) (1976). *Advanced Geostatistics in the Mining Industry*, Volume 24 of *NATO Advanced Study Institute Series; Series C: Mathematical and Physical Sciences*. Dordrecht: D. Reidel Publishing Company. 461 p.

Helterbrand, J. D. and N. Cressie (1994). Universal cokriging under intrinsic coregionalization. *Mathematical Geology 26*(2), 205–226.

Hendriks, L. A. M., H. Leummens, A. Stein, and P. J. Debruijn (1998). Use of soft data in a GIS to improve estimation of the volume of contaminated soil. *Water, Air and Soil Pollution 101*(1–4), 217–234.

Isaaks, E. H. and R. M. Srivastava (1989). *An Introduction to Applied Geostatistics*. New York: Oxford University Press. 561 p.

Jobson, J. D. (1992). *Applied Multivariate Data Analysis. Categorical and Multivariate Analysis*. Springer Texts in Statistics. Volume II. New York: Wiley–Interscience. 731 p.

Johnson, N. L., S. Kotz, and N. Balakrishnan (1994). *Continuous Univariate Distributions,* 2 ed. Wiley Series in Probability and Statistics, 2 Vols. New York: Wiley-Interscience. 1536 p.

Journel, A. G. (1977). Kriging in terms of projections. *Mathematical Geology 69*(6), 563–586.

Journel, A. G. (1980). The lognormal approach to predicting local distributions of selective mining unit grades. *Mathematical Geology 12*(4), 285–303.

Journel, A. G. (1998). Markov models for cross covariances. Technical Report 11, Stanford Center for Reservoir Forecasting, Stanford, Calif. 13 p.

Journel, A. G. and C. J. Huijbregts (1978). *Mining Geostatistics*. London: Academic Press. 600 p.

Kemény, P. (1984). Grundbegriffe der Matrix-Algebra. See Fahrmeir and Hamerle (1984), pp. 689–723.

Kendall, M. and A. Stuart (1979). *The Advanced Theory of Statistics. Volume II: Inference and Relationship*. London: Charles Griffin & Co., Ltd. 748 p.

Kendall, M., A. Stuart, and J. K. Ord (1983). *The Advanced Theory of Statistics*. Vol. III: *Design and Analysis, and Time-Series*. London: Charles Griffin & Co., Ltd. 780 p.

Kirsch, C. and V. Pawlowsky (1985). Modélisation de variogrammes croisés: comment satisfaire à l'inégalité de Cauchy-Schwarz. *Science de la Terre, Ser. Informatique 24*, 53–62.

Kotz, S., N. Balakrishnan, and N. L. Johnson (2000). *Continuous Multivariate Distributions*. Vol. I. *Models and Applications*. Wiley Series in Probability and Statistics. New York: Wiley-Interscience. 730 p.

Krajewski, W. F. (1987). Cokriging radar-rainfall and rain gauge data. *Journal of Geophysical Research 92*(8), 9571–9580.

Kres, H. (1983). *Statistical Tables for Multivariate Analysis.* New York: Springer-Verlag. 504 p.

Krige, D. G. (1978). *Lognormal-de Wijsian Geostatistics for Ore Evaluation.* Johannesburg: South African Institute of Mining and Metallurgy. 51 p.

Krzanowski, W. J. (1988). *Principles of Multivariate Analysis: A User's Perspective.* Oxford Statistical Science Series. Volume III. Oxford: Clarendon Press. 563 p.

Leenaers, H., J. P. Okx, and P. A. Burrough (1989). Co-kriging: An accurate and inexpensive means of mapping floodplain soil pollution by using elevation data. See Armstrong (1989), pp. 371–382.

Lippard, S. J., A. Næss, and R. Sinding-Larsen (Eds.) (1999). *Proceedings of IAMG'99—The Fifth Annual Conference of the International Association for Mathematical Geology.* Trondheim: Tapir. 2 Vols., 784 p.

Ma, X. and T. Yao (2001). A program for 2d modeling (cross) correlogram tables using Fast Fourier Transform. *Computers & Geosciences 27*(7), 763–774.

Martín-Fernández, J. A., C. Barceló-Vidal, and V. Pawlowsky-Glahn (1997). Different classifications of the Darss Sill data set based on mixture models for compositional data. See Pawlowsky-Glahn (1997), pp. 151–156.

Martín-Fernández, J. A., C. Barceló-Vidal, and V. Pawlowsky-Glahn (1998a). Measures of difference for compositional data and hierarchical clustering methods. See Buccianti, Nardi, and Potenza (1998), pp. 526–531.

Martín-Fernández, J. A., C. Barceló-Vidal, and V. Pawlowsky-Glahn (1998b). A critical approach to non-parametric classification of compositional data. In A. Rizzi, M. Vichi, and H.-H. Bock (Eds.), *Advances in Data Science and Classification (Proceedings of the 6th IFCS Conference—IFCS'98)*, pp. 49–56. Università "La Sapienza," Rome. International Federation of Classification Societies. Berlin: Springer-Verlag.

Martín-Fernández, J. A., C. Barceló-Vidal, and V. Pawlowsky-Glahn (2000). Zero replacement in compositional data sets. In H. Kiers, J. Rasson, P. Groenen, and M. Shader (Eds.), *Studies in Classification, Data Analysis, and Knowledge Organization (Proceedings of the 7th IFCS Conference—IFCS2000)*, pp. 155–160. University of Namur, Belgium. International Federation of Classification Societies. Berlin: Springer-Verlag.

Martín-Fernández, J. A., M. Bren, C. Barceló-Vidal, and V. Pawlowsky-Glahn (1999). A measure of difference for compositional data based on measures of divergence. See Lippard, Næss, and Sinding-Larsen (1999), pp. 211–216.

Mateu-Figueras, G., C. Barceló-Vidal, and V. Pawlowsky-Glahn (1998). Modeling compositional data with multivariate skew-normal distributions. See Buccianti, Nardi, and Potenza (1998), pp. 532–537.

Matheron, G. (1965). *Les Variables Régionalisées et leur Estimation—Une Application de la Théorie des Fonctions Aléatoires aux Sciences de la Nature.* Paris: Masson et Cie. 305 p.

Matheron, G. (1971). The theory of regionalized variables and its applications. Technical Report C–5, École Nationale Supérieure des Mines de Paris, Centre de Geostatistique et de Morphologie Mathematique, Fontainebleau. 211 p.

Matheron, G. (1979). Recherche de simplification dans un probleme de cokrigeage. Technical Report N–698, École Nationale Supérieure des Mines de Paris, Centre de Geostatistique et de Morphologie Mathematique, Fontainebleau. 19 p.

McBratney, A. B. and R. Webster (1983). Optimal interpolation and isarithmic mapping of soil properties. *Journal of Soil Science 34*, 137–162.

Mood, A. M. and F. A. Graybill (1963). *Introduction to the Theory of Statistics.* New York: McGraw Hill. 443 p.

Mood, A. M., F. A. Graybill, and D. C. Boes (1986). *Introduction to the Theory of Statistics* (17 ed.). New York: McGraw-Hill. 564 p.

Morisette, J. (1997). Examples using SAS to fit the model of linear coregionalization. *Computers & Geosciences 23*(3), 317–323.

Muge, F. H. and G. Cabeçadas (1989). A geostatistical approach to eutrophication modelling. See Armstrong (1989), pp. 445–457.

Myers, D. E. (1982). Matrix formulation of co-kriging. *Mathematical Geology 14*(3), 249–257.

Myers, D. E. and J. Carr (1984). Co-kriging and principal component analysis. Bentonite data revised. *Science de la Terre, Sér. Informatique 21*, 65–77.

Natke, H. G. (1983). *Einführung in Theorie und Praxis der Zeitreihen- und Modalanalyse.* Braunschweig, Germany: Friedrich Vieweg & Sohn. 522 p.

Olea, R. A. (Ed.) (1991). *Geostatistical Glossary and Multilingual Dictionary.* New York: Oxford University Press. 177 p.

Olea, R. A. (1999). *Geostatistics for Engineers and Earth Scientists.* Norwell, Mass.: Kluwer Academic Publishers. 303 p.

Olea, R. A., V. Pawlowsky, and J. C. Davis (1993). Volumetric calculations in an oil field: The basis method. *Computers & Geosciences 19*(10), 1517–1527.

Olea, R. A. and V. Pawlowsky-Glahn (1999). Geostatistical analysis and characterization of compositional data from an oil field in Kansas, USA. See Lippard, Næss, and Sinding-Larsen (1999), pp. 499–504.

Pan, G. (1993). Regionalized favorability theory for information synthesis in mineral exploration. *Mathematical Geology 25*(5), 603–631.

Pan, G., D. Gaard, K. Moss, and T. Heiner (1993). A comparison between cokriging and ordinary kriging: Case study with a polymetallic deposit. *Mathematical Geology 25*(3), 377–398.

Papoulis, A. (1965). *Probability, Random Variables and Stochastic Processes.* New York: McGraw-Hill. 583 p.

Pawlowsky, V. (1984). On spurious spatial covariance between variables of constant sum. *Science de la Terre, Sér. Informatique 21*, 107–113.

Pawlowsky, V. (1986). *Räumliche Strukturanalyse und Schätzung ortsabhängiger Kompositionen mit Anwendungsbeispielen aus der Geologie.* Ph. D. thesis, Fachbereich Geowissenschaften, Freie Universität Berlin, Germany. 170 p.

Pawlowsky, V. (1989). Cokriging of regionalized compositions. *Mathematical Geology 21*(5), 513–521.

Pawlowsky, V. and H. Burger (1992). Spatial structure analysis of regionalized compositions. *Mathematical Geology 24*(6), 675–691.

Pawlowsky, V., R. Olea, and J. C. Davis (1995). Estimation of regionalized compositions: A comparison of three methods. *Mathematical Geology 27*(1), 105–127.

Pawlowsky, V. and K. Stattegger (1988). Cokriging of compositional data—A case study from modern stream sands. *Science de la Terre, Sér. Informatique 28*, 267–280.

Pawlowsky-Glahn, V. (Ed.) (1997). *Proceedings of IAMG'97—The Third Annual Conference of the International Association for Mathematical Geology.* Barcelona: International Center for Numerical Methods in Engineering (CIMNE). 2 Vols. and Addendum, 1100 p.

Pearson, K. (1897). Mathematical contributions to the theory of evolution: On a form of spurious correlation which may arise when indices are used in the measurement of organs. *Proceedings of the Royal Society of London LX*, 489–502.

Pereira, H. G., A. O., and Soares (1989). Application of geostatistics to groundfish survey data. See Armstrong (1989), pp. 459–467.

Pesti, G., I. Bogardi, W. E. Kelly, and R. J. Kalinski (1993). Cokriging of geoelectric and well data to define aquifer protective layers. *Ground Water 31*(6), 905–912.

Pukelsheim, F. (1994). The three sigma rule. *The American Statistician 48*(2), 88–91.

Quintana, J. M. and M. West (1988). Time series analysis of compositional data. In J. M. Bernardo, M. H. DeGroot, D. V. Lindley, and A. F. M. Smith (Eds.), *Bayesian Statistics 3*, pp. 747–756. New York: Oxford University Press.

Rayens, W. S. and C. Srinivasan (1991). Box–Cox transformations in the analysis of compositional data. *Journal of Chemometrics 5*, 227–239.

Rendu, J.-M. (1978). *An Introduction to Geostatistical Methods of Mineral Evaluation.* Johannesburg: South African Institute for Mining and Metallurgy. 84 p.

Rendu, J.-M. (1979). Normal and lognormal estimation. *Mathematical Geology 11*(4), 407–422.

Ribeiro, L. T. and F. H. Muge (1989). A geostatistical approach to the modelling of a piezometric field. See Armstrong (1989), pp. 651–660.

Rivoirard, J. (1990). A review of lognormal estimators for *in situ* reserves (teacher's aid). *Mathematical Geology 22*(2), 213–221.

Royer, J. J. (1989). Multivariate geostatistics and sampling problems. See Armstrong (1989), pp. 823–836.

Samper-Calvete, F. J. and J. Carrera-Ramírez (1990). *Geoestadística—Aplicaciones a la Hidrología Subterránea.* Barcelona: Centro Internacional de Métodos Numéricos en Ingeniería (CIMNE). 484 p.

Sanford, R. F., C. T. Pierson, and R. A. Crovelli (1993). An objective replacement method for censored geochemical data. *Mathematical Geology 25*(1), 59–80.

Smyth, J. D. and J. D. Istok (1989). Multivariate geostatistical analysis of groundwater contamination by pesticide and nitrate: A case history. See Armstrong (1989), pp. 713–724.

Solow, A. R. and S. M. Gorelick (1986). Estimating monthly stream flow values by cokriging. *Mathematical Geology 18*(8), 785–809.

Sousa, A. J. (1989). Geostatistical data analysis—An application to ore typology. See Armstrong (1989), pp. 851–860.

Stattegger, K. (1983). Application of time series analysis to the tectonic analysis of disturbed rock sequences recorded from drill hole logs with examples from the paleozoic of Graz (Austria). *Mathematical Geology 15*(6), 673–685.

Stein, A. and L. Corsten (1991). Universal kriging and cokriging as a regression procedure. *Biometrics 47*, 575–587.

Stein, A., A. van Eijnsberger, and L. G. Barendregt (1991). Cokriging nonstationary data. *Mathematical Geology 23*(5), 703–719.

Stuart, A. and J. K. Ord (1987). *Kendall's Advanced Theory of Statistics,* Volume 1: *Distribution Theory.* New York: Oxford University Press. 604 p.

Tercan, A. E. and P. A. Dowd (1995). Approximate local confidence intervals under change of support. *Mathematical Geology 27*(1), 149–172.

Trangmar, B. B., R. S. Yost, and G. Uehara (1986). Spatial dependence and interpolation of soil properties in West Sumatra, Indonesia: II. Co-regionalization and co-kriging. *Soil Science Society of America Journal 50*, 1396–1400.

Unal, A. and C. Haycocks (1986). Geostatistical coal resource characterization using intervariable correlations. In R. V. Ramani (Ed.), *Application of Computers and Operations Research in the Mineral Industry 19*, pp. 231–239. Littleton, Colo.: Society of Mining Engineers, Inc.

ver Hoef, J. M. and N. Cressie (1993). Multivariate spatial prediction. *Mathematical Geology 25*(2), 219–240.

Wackernagel, H. (1998). *Multivariate Geostatistics: An Introduction with Applications* (2 ed.). Berlin: Springer-Verlag. 291 p.

Wackernagel, H. and C. Butenuth (1989). Caractérisation d'anomalies géochimiques par la géostatistique multivariable. *Journal of Geochemical Exploration 32*, 437–444.

Yaglom, A. M. (1986). *Correlation Theory of Stationary and Related Random Functions.* I: *Basic Results.* II: *Supplementary Notes and References.* New York: Springer-Verlag. 526 p.

Yao, T. (1998). *Automatic covariance modeling and conditional spectral simulation with Fast Fourier Transform.* Ph. D. thesis, Department of Geological and Environmental Sciences, School of Earth Sciences, Stanford University, Stanford, Calif. 173 p.

Yao, T. and A. G. Journel (1998). Automatic modeling of (cross) covariance tables using Fast Fourier Transform. *Mathematical Geology 30*(6), 589–615.

Yates, S. R. and A. W. Warrick (1987). Estimating soil water content using cokriging. *Soil Science Society of America Journal 51*(1), 23–30.

Index